FISH ENDOCRINOLOGY

ENCYCLOPAEDIA OF ENDOCRINOLOGY-I

FISH ENDOCRINOLOGY

By

Manju Yadav

Lecturer

Department of Zoology

M.M.H. College

Ghaziabad (U.P.)

(India)

DISCOVERY PUBLISHING HOUSE PVT. LTD.

NEW DELHI-110 002

First Published-2008

Reprinted-2014

ISBN: 978-93-5056-570-4 (Set)

ISBN: 978-81-8356-271-3

Published by:

DISCOVERY PUBLISHING HOUSE PVT. LTD.
4383/4B, Ansari Road, Darya Ganj
New Delhi-110 002 (India)
Phone: +91-11-23279245, 43596064-65
Fax: +91-11-23253475
E-mail: discoverypublishinghouse@gmail.com
sales@discoverypublishinggroup.com
web: www.discoverypublishinggroup.com

Printed at:
Infinity Imaging Systems
Delhi

Preface

The present title **Fish Endocrinology** provides a definitive and comprehensive review of all aspects relating to hormones in different animals including wild and domestic species. It discusses the intimate physiology of the endocrine system itself and describes the role of hormones in the processes of nutrition, osmoregulation, colour change, calcium metabolism and reproduction. Subject matter has been presented in a readable way emphasizing the differences between species from a functional aspects. It not only contains a wealth of information concerning the endocrinology of the various species but also contains more than enough basic information concerning the biosynthesis, metabolism and mode of action of the hormones for the book to be read and understood without referring to a text book of basic endocrinology. This book has been written primarily for use as a text book by under-graduate, as well as graduate students.

Efforts have also been made to make the presentation lucid and accessible for beginning students, emphasizing major concepts while in corporating the latest experimental findings and theories throughout. The aim is to enthuse the readers with this active and exciting area of research and to lay a solid foundation on which further study of its various facts may be based.

Though, the author has taken special care to present a current account, yet he is fully aware about limitations and the readers may come across the mistakes of various types. For all types of mistakes author extends his due apology.

There can be no claim to originality except in the manner of treatment and much of the information has been obtained from the books and scientific journals available in the different libraries.

The author expresses his thanks to his friends and colleagues whose continue inspirations have initiated him to bring out this book.

The author expresses his gratitude to Mr. Wasan and staff of M/s Discovery Publishing House for their whole hearted co-operation in the publication of this book.

Author

CONTENTS

1

PINEAL GLAND

Despite the vast number of publications on the pineal body of fishes, no distinct function has been unequivocally demonstrated. It is still impossible to present a unified concept of pineal function; rather an attempt will be made to review some of the evidence for the various current hypotheses concerning its function.

PINEAL DILEMMA

The most persistent question concerning the pineal gland of fishes is whether it is the vestigial remnant of an ancestral third eye or whether it is a specialized functional organ. Its location on the top of the head orients the pineal organ directly toward the primary source of light and thereby effectively eliminates the pineal body as a receptor of a visual field. But this orientation renders the pineal well suited for detecting changes in light intensity or for acting as a dosimeter of incident radiation. In other words, the pineal of fishes probably never was, in the true sense, an eye. Nevertheless, numerous investigators have shown that the pineal organ of the fishes does have a light receptive function. The problem then is not the presence or absence of a function, but rather, a question of the precise function or functions.

Although several of the current hypotheses are not mutually exclusive, they can, for present purposes, be divided into two categories—one emphasizing primarily a sensory role and the other a secretory function. Of the sensory theories, the following are the most credible: (1) The pineal body is a photosensory structure, (2) it acts as a baro- or chemoreceptor for the cerebrospinal fluid (CSF), or, (3) it performs the function of a mediator in the olfactory induced response to exohormones. Conversely, (4) the pineal may be primarily a gland of external secretion related to the chemical composition of the CSF or the metabolism of the brain tissue, or (5) it may be a gland of internal secretion with an endocrine function.

It is not proposed here to review all of the literature concerning these various hypotheses but to provide a framework into which the accumulated knowledge of pineal function in the fishes can be channeled.

Ontogeny and Structure of the Pineal Organ

The description of pineal ontogeny, morphology, and histology given here may be amplified from detailed accounts.

Ontogeny

Much of the polemic surrounding the pineal organ of the fishes originates from the prevailing ignorance of its embryological origin. This has given rise to inconsistent nomenclature and a failure to distinguish clearly the pineal body from other epithalamic outgrowths.

In vertebrates, two approximately middorsal evaginations develop from the roof of the diencephalon. The rostral evagination represents the anlage of the parapineal body which is homologous with the parietal eye of certain reptiles. The pineal organ develops from the more posterior of the diencephalic evaginations. Therefore, the parapineal and pineal organs are two independent structures and are not homologous. The paraphysis, which has occasionally been confused with the parapineal, is morphologically part of the telencephalon and is not part of the pineal area.

In fishes, the parapineal anlage usually shows some early ontogenetic development but is generally absent or rudimentary in adults, including the primitive ganoid *Amia calva* and the dipnoan *Protopterus annectens*. However, a parapineal body resembling a parietal eye of lizards was described in at least one adult teleost. In other groups of fishes, a well-developed parapineal body with distinguishable photosensory-like cells is found only among the nonmyxinoid cyclostomes. However, the same author notes that the problem of the parapineal organ is not resolved and should be completely reexamined.

Structure

Topography and general morphology

The pineal body is a conspicuous structure in most adult fishes; it is, however, rudimentary in *Syngnathus acus* and *Hippocampus spinosa* and absent in *Torpedo ocellata*, *T. marmorata* (Selachii), and the adult *Myxine glutinosa*. Despite contradictory opinions, a pineal organ is present in the tuna *Thynnus thynnus*.

In basic form, the fish pineal organ consists of a hollow, variously invaginated and well-vascularized structure lying dorsal to the diencephalon to which it is connected by a hollow narrow stalk. The pineal body consists of an expanded distal vesicular area containing a central lumen which connects with the third ventricle via the narrow proximal stalk. Posteriomedially the stalk connects with the subcommissural organ and the commissura posterior and makes contact with the commissura habenulae on its anterior side before joining the dorsal sac epithelium which comes to lie directly ventral to the distal vesicle of the pineal body. With advancing age, the epithelium of the distal vesicle undergoes a gradual infolding with the result that the central lumen, although retaining its connection with the third ventricle, becomes divided into narrow sinuous space's. Many of the blood vessels which surround the pineal body in embryonic and larval stages, are, with the increased development of the animal, carried into the infoldings of the pineal epithelium and come to lie in the intralobular septa. This centripetal invasion by the blood vessels results in the extensive vascularization characteristic of the adult pineal. Generally, the distal vesicle is closely apposed to the overlying cranial roof but is not, in teleosts, associated with a pineal foramen in the skull. However, as in many deep-sea fishes, the pineal body often lies beneath a more or less depigmented and translucent area of the cranial roof. In other fish, the pineal body is located in a depression in the roof of the skull or may be associated with a very complex structure of skull bones which may allow the passage of light directly to the pineal body. Some authors, in questioning a photosensory role of the pineal body, have attached special significance to the fact that some fish possess thick skull bones and have no special structure to allow light to pass though to the pineal region. It should be noted, however, that measurable quantities of light have been demonstrated to penetrate the skulls of sheep, dogs, rabbits, and rats.

Histology

Histologically, the pineal body of fishes most closely resembles a sensory structure. The structure of the vesicular epithelium resembles the retina except that the pineal "retina" is of a reversed type with the sensory cells in the inner retinal layer; thus, they are turned in the direction from which the light enters. In cyclostomes the dorsal wall of the pineal forms a lenslike pellucida, but a lenslike differentiation is never found among other fishes where the pineal seems to have acquired some characteristics of a secretory organ.

Despite this histological progression from an eyelike structure in some fish to a glandular organ in others, there is now general agreement that the pineal body of fish is largely composed of three cell types: sensory cells, supporting cells, and ganglion or nerve cells.

Most of the pineal cells are sensory cells. These are most abundant in the vesicular area on the side of the lumen but occur throughout the pineal epithelium. These cells are large, ovoid or club-shaped with well-defined nuclei surrounded by clear cytoplasm containing many mitochondria and capped with a membranous structure—an outer segment. Cytoplasmic outgrowths arising from the sensory cells and extending into the pineal lumen have also been described. On the pole to the cytoplasmic outgrowths, the sensory cells give off axons, which together with the dendrites from the ganglion cells, form a network of axons and dendrites within the pineal organ.

Supporting cells of various types show a marked affinity for nuclear stains and lack the typical structure of the sensory cells. It is most likely that these cells are of an ependymal glia type.

Ganglion or nerve cells together with the dendrites and Nissl substance indicative of their nervous nature, have been described in the pineal of some but not all fishes examined. Although few in number compared with the other cell types, ganglion cells may be readily distinguished by their large size and by the presence of two or three well-stained nucleoli. Recently, Rudeberg (1968b) has reported that within the pineal of the sardine, *Sordina pilchardus sardina*, the ganglion cells are limited to three small groups. If this localization of ganglion cells proves to be the case in other fishes, it may account for the failure of some workers to find pineal ganglion or nerve cells.

Innervation

Apart from the sensory cells proper, afferent nerve fibers connecting the pineal body with other parts of the brain have been described. These nerves, which appear to represent the axons of ganglion cells, arise from the extensive nerve net formed within the pineal by the dendrites of the ganglion cells and neuraxis of the primary sensory cells. In the stalk region, these fibers form variable numbers of nerve bundles before reaching the roof of the diencephalon. Efferent fibers originating within the central nervous system and ending on pineal cells have rarely been described. These efferent fibers are probably the same type of aberrant commissural fibers described in mammals.

Aside from a well-documented nervous connection between the pineal body and the posterior commissure relatively little is known of the central terminations of the pineal nerve. Possible connections have been suggested with the superior commissure, and with the right habenular nucleus as well as other habenular structures. In one case only, fibers have been described as running to the optic tectum. Some workers have suggested a link between the pineal body and the subcommissural organ, but this is apparently not present in all fishes. Le Gros Clarke (1932) believes that the pineal organ of cyclostomes is heavily supplied with olfactory input channels and this possibility should be investigated. Notwithstanding the paucity of information concerning the exact site(s) of the pineal nerve terminations, connections already described could permit a link between the pineal body and the efferent centers of the brain through which the pineal body could influence various integrative afferent centers.

Physiology of the Pineal Body

Although neurophysiological studies have proved that the pineal organ of *Salmo gairdneiri irideus* is responsive to illumination there is only scanty literature on the actual physiology of the pineal organ in fishes. Active cellular metabolism has been indicated by at least two authors who studied the uptake of radioactive phosphorus by the fish pineal.

Recently, interest has been directed toward the possible presence of melatonin and other tryptophan derivatives within the fish pineal. Quay (1965) localized the enzyme hydroxyindole-*O*-methyl transferase (HIOMT) which is responsible for the formation of melatonin within the pineal of *Salmo irideus*. Working with the same species, Oguri et al. (1968) found that a melatonin precursor, 5-hydroxytryptophan is taken up by the pineal organ in greater quantities than it is by other parts of the brain studied; they infer from this finding that the pineal of fish may also be concerned with melatonin synthesis. These studies have now been substantiated by Fenwick (1970a) who isolated melatonin from the pineal organs of the Pacific salmon *Oncorhyncus tshawytscha*. Thus, the pineal organ of fish is responsive to light, has a high cellular metabolism, and shows an active tryptophan metabolism which is capable of producing the mammalian hormone melatonin.

Pineal Body as a Sensory Organ

There remains little doubt that the pineal body of some, if not all fish, functions as a sensory organ. Krabbe (1916) and Walter (1923) and more recently Van de Kamer (1952) as well as Hafeez and Ford (1967) have suggested that the fish pineal is associated with the detection of the pressure or chemical composition of the CSF. Evidence favouring this hypothesis includes the reported open connection between the pineal lumen and the ventricles of the brain, the changing shape of the pineal body of *Esox lucius* in relation to the pressure of the CSF, and the cytoplasmic extensions of the primary sensory cells toward the lumen of the body thus suggesting a functional relationship with the CSF. Hafeez and Ford (1967) go on to suggest that the similarities between the pineal body and the subcommissural organ with regard to the chemical nature of their secretions and the common direction of release into the ventricles support the view of a CSF-pineal relationship. There have been no reported studies of the effect of pinealectomy on either the pressure or composition of the CSF.

Although the hypothesis that the pineal gland is associated with the limbic system or visceral brain suffers from the same paucity of experimental evidence, some suggestive findings have been reported. Both Le Gros Clarke (1932) and Boon (1938) noted a close anatomical relationship between the pineal body and the olfactory system of fishes; the latter has been shown to have at least some connection with hypothalamic nuclei responsible for pituitary hormone release. Moreover, Hoffman and Reiter (1965) reported that olfactory stimuli are capable of interfering with the primary inhibitory effects which the active pineal gland has on the reproductive system of male hamsters. Such studies have not yet been extended to fishes.

In contrast to the scanty evidence for these pineal functions, many publications suggest a photosensory role as the primary sensory function of the pineal organ. The evidence, based on a number of fishes, comes from direct neurophysiological studies as well as indirect evidence from histology or ultrastructure, colour changes, and behavioural tests.

By light as well as electron microscopic investigation, it has been clearly demonstrated that morphologically the neurosensory cells present in the epithelium of the pineal body are very similar to the ciliary type of photosensory cells present in the retina of the lateral eyes and possess characteristics which are indisputably conelike. It has been suggested by De la Motte (1963) that the light receptive role of the pineal body in *Phoxinus* is based on porphyropsin, although Thines and Kahling (1957) had previously reported a different visual pigment in *Anoptichthys*. More directly, Grunewald-Lowenstein (1956) observed histological and histochemical changes in the pineal organ of *Astanax* following prolonged exposure to continuous darkness or illumination. Similar changes, however, were not found in young sockeye salmon held under different light regimes. Recently, Dodt (1963) and Morita

(1966) have demonstrated the photosensory potential of the fish pineal by electrophysiologically detecting alterations in nervous activity in the pineal of *Salmo irideus* during and after illumination of the pineal even though this species lacks any specialized tissue overlying the pineal organ.. This finding is important when attempting to define a light receptive role for the pineal in those fishes which possess apparently light impermeable skulls. It questions the theory that such a skull precludes a photosensory role.

A light receptive role of the pineal organ is also indicated by light-induced alterations of pigment distribution within the chromatophores. Although several studies suggested that the pineal had no effect on pigment distribution, von Frisch did find that some part of the diencephalon was involved in chromatophore responses. The pineal was more directly implicated by Young (1935) who found that pineal extirpation abolished the marked diurnal rhythm of colour change in ammocoetes and distributed the rhythm in adult *Lampetra*. Subsequently, Breder and Rasquin (1950), Hoar (1955), and Schonherr (1955) reported some degree of pigmentary dispersion after pineal occlusion or destruction—a finding similar to that reported by Young (1935) in pinealectomized ammocoetes. These results are supported by the observation that administration of beef pineal extracts in embryonic and larval *Fundulus* caused marked pallor although similar results could not be obtained in adult *Fundulus* or adult *Phoxinus*. The difference in responsiveness between the young and adults was presumed by Fain and Hadley (1966) to result from the acquisition, in the adults, of nervous innervation mediated by catecholamines and a loss of sensitivity to the pineal extracts. Whether the pallor following pineal extract administration is the result of pituitary inhibition or a direct action on the pigment cells is not known. A direct action is, however, suggested by the ability of the pineal extract to cause pigment concentration *in vitro*.

Photosensitivity of the pineal body has been indirectly demonstrated in several species of fish by the use of behavioural tests. Breder and Rasquin (1947), Hoar (1955), and Fenwick (1970b) report that phototaxis is abolished following pinealectomy. Further, Breder and Rasquin (1947) and Fenwick (1970b) found that phototaxis, whether positive or negative, depended on the presence of intact optic cysts or intact eyes; they concluded that although the sign of phototaxis was governed by the pineal organ, the phenomenon itself depended on the presence of lateral photic receptors. Hoar (1955), however, working with young sockeye salmon smolts, *Oncorhynchus nerkri*, found that the negative phototaxis of otherwise intact animals is not disturbed following damage to the pineal region; this difference may result from species or age differences of the animals or differences in the intensity of light employed in the experiments. The pineal organ of fishes becomes increasingly invaginated with age and undergoes a decrease in sensory cell number with continued development together with a marked degeneration of nervous elements. Thus, the pineal body of fishes may change from a primary photosensory structure in the young fish to a secondary photosensory structure in the older fish where it can no longer autonomously produce a phototactic response. On the other hand, Hoar (1955) performed his experiments outside where the light intensities were much greater than those employed in other studies. Previously, von Frisch (1911a) and Young (1935) suggested a general light sensitivity of the diencephalic roof; thus, since the smolts used by Hoar (1955) had thin skulls, the phototaxis demonstrated in "pinealectomized" fish could have resulted from the effect of high light intensity on the brain itself.

Recent studies by Fenwick (1970b) have shown that intact goldfish in a light gradient were unevenly distributed and spent most of their time in the darker half of the tank. Conversely, pinealectomized, bilaterally enucleated, or pinealectomized plus bilaterally enucleated goldfish were distributed uniformly throughout the gradient. From this evidence it was concluded that the phototactic response of goldfish depends upon the presence of the pineal organ as well as the eyes. Furthermore, in a conditioning situation, pinealectomized animals with intact vision showed significantly more responses to the conditioned stimulus than did the controls when the conditioned stimulus was light, but not when the conditioned stimulus was sound. Blind goldfish, with or without an intact pineal, could not be effectively conditioned to light although they did become conditioned to sound. From this data it was concluded that although the pineal organ of goldfish is a photosensory organ, in the absence of the eyes, the photic information received by the pineal cannot be translated into a directional response or be used as a sensory mechanism for the initiation of active behaviour. It appears, therefore, that the photosensory role of the goldfish pineal organ is to modulate the response elicited by photic information received by the eyes.

Although these studies do not provide conclusive evidence of the mode of action or the exact role of the pineal body in light reception, they do demonstrate the importance of the pineal body in phototactic behaviour and suggest that the pineal organ and the eyes function as a unit in phototaxis.

Pineal Body as a Secretory Organ

External secretion

In addition to its photosensory role, many investigators have shown that fish pineal produces an apocrine secretion which contains glycogen. The secretion is generally considered to enter the CSF. Although Van de Kamer (1955) suggests that the secretory droplets may be experimental artifacts, the high cellular metabolism demonstrated by the cytological studies of Palayer (1958) and U. Holmgren (1959a) and the presence of a well-developed Golgi complex in the supporting cells and the sensory cells indicate secretory activity.

Internal secretion

There is also the possibility of an internal secretory or endocrine role. Although both Friedrich-Freksa (1932) and U. Holmgren (1959a) suggest that some pineal secretion is taken up by the blood vessels in teleosts, Hafeez and Ford (1967) did not find any secretory granules within the nerve fibers or in the proximity of the blood vessels of sockeye salmon, *Oncorhynchus nerka*. However, since the demonstration of secretory granules depends upon the nature of the secretion and on the specificity of the techniques employed, an internal secretory role of the pineal organ was not definitely ruled out and few investigators have applied techniques suitable for detecting the one known mammalian pineal hormone melatonin.

Melatonin and its precursor serotonin, together with the enzyme HIOMT necessary for the formation of melatonin, have now been demonstrated in fish pineal organs. Also, Oguri et al. (1968) report that the pineal takes up more ^{14}C-5-hydroxytryptophan than any other tissues studied. Since this substance is a precursor of serotonin, an active tryptophan metabolism within the fish pineal organ is indicated. Thus, an endocrine role of the fish pineal, possibly based on the hormone melatonin, is not improbable.

Despite earlier reports by Krockert (1936a,b) that the growth rate of young *Lebistes* could be decreased by feeding them desiccated bull pineal glands, Pflugfelder (1953, 1954, 1956a) clearly implicated the pineal as an endocrine organ of fishes and suggested a pineal-pituitary relationship in teleosts. His observations on pinealectomized guppies indicated hypertrophy of the adenohypophysis, hyperthyroidism, decreased growth rate, a slight acceleration in the appearance of secondary sex characteristics in young males, an increase in the activity of the interrenal cells, and a disturbed calcium metabolism resulting in spinal curvature. The thyroidal effects could be partially offset by the injection of epiphysan or thyroxin. Pflugfelder (1964) has since reported alterations in the pars distalis and thyroid in pinealectomized goldfish. U. Holmgren (1959b) reported similar findings with regard to skeletal abnormalities and was able to reduce these by administering beef pineal injections. Further, the latter investigator reported a decreased radiocalcium uptake in pinealectomized fish and an increased uptake in those fish receiving injections of pineal extract. These findings, however, could not be confirmed by Weisbart and Fenwick (1966) who report an undisturbed blood calcium level in pinealectomized goldfish.

Few other studies have examined the endocrine nature of the fish pineal, and these have been conflicting. Contrary to the report of thyroidal hypertrophy by Pflugfelder (1964), Pang (1967) found a decreased thyroidal cell height in pinealectomized *Fundulus* and Rasquin (1958), U. Holmgren (1959b), Fenwick (1970c), and Peter (1968) report no alteration in this tissue following pinealectomy. Further, neither Peter (1968) nor Fenwick (1970c) could find any apparent increase in the interrenal cells as previously reported by Pflugfelder (1953). While the significance of these findings is not clear, the possibility that the pineal body of mammals is related to electrolyte balance has never been wholly repudiated and should provide the impetus for further investigation in this area.

Although a pineal-gonadal relationship seems well established in the mammals, little research has been carried out on the possibility of such an axis in the fishes. Krockert (1936a,b) was able to delay the appearance of secondary sex characteristics in young *Lebistes* by feeding desiccated bull pineal glands. In the same species, Pflugfelder (1954) reports a slight acceleration in the sexual development of males following pinealectomy. In contrast to Pflugfelders results, Schonherr (1955) and Rasquin (1958) reported unsuccessful attempts to hasten reproductive maturation by pinealectomy. Pang (1967) described a delay in the appearance of nuptial colouration in pinealectomized *Fundulus*, but noted that the controls had smaller gonad sizes. It is not clear whether this means that his controls underwent gonadal atrophy or that his pinealectomized fish showed gonadal hypertrophy. Recently, Peter (1968) and Fenwick (1970c), both working with goldfish, have reported contradictory results. Peter (1968), in one experiment of 9-

months duration, could find no significant differences between the gonosomatic indices (GSI) (gonad weight/body weight × 100 = GSI) of pinealectomized, sham-pinealectomized, and unoperated control goldfish. However, when experiments were carried out at different stages of the yearly reproductive cycle, Fenwick (1970c) found that although pinealectomy had no effect on the size of the gonads during most of the year, the operation, when carried out just prior to the onset of the final maturation phase preceding the normal spawning period, did result in a highly significant increase in the GSI relative to that found in the groups with intact pineal. The time at which pinealectomy was found to have this effect corresponded with that period of the reproductive cycle during which an increasing day length had its greatest stimulatory effect on the reproductive system. Further, the increase in gonad size in goldfish subjected to increasing periods of daily light exposure could be prevented by the daily injection of melatonin.

These results may help to explain the earlier contradictory reports. In the first place it seems probable that the pineal gland may not function to a similar extent, or even in an identical manner, during different periods of the life cycle. As seen previously, the invagination of the pineal gland increases with age and there is a decrease in sensory cell number together with a marked degeneration of nervous structures. Further, the pineal gland becomes less accessible to light with increase in body size; even the effect of pinealectomy on pigment distribution differed in young and older lampreys. Second, if the endocrine function of the pineal organ is dependent upon melatonin as suggested by Fenwick (1970a,c), and if the level of melatonin is depressed by exposure to long periods of bright light as reported in the mammals, then the experimental design must be considered before interpreting the results of pinealectomy. If such controls are not carefully enforced, the intensity and duration of the experimental photoperiod may be sufficient to lower the melatonin level to a point where even animals with a pineal are effectively "physiologically pinealectomized."

Pineal Body as a Sensory cum Secretory Organ

The apparent disagreement between workers supporting a sensory function as opposed to those who argue for a secretory function does not seem justified. Rather, it seems likely that both functions are linked and that the pineal organ should be viewed neither as an entirely sensory structure nor as a wholly secretory gland; both sensory and secretory functions seem likely. This hypothesis is supported by recent observations made on the chelonian epiphysis. Vivien and Roels (1967) noted that certain cells within the chelonian epiphysis could be seen to alternate between characteristics of reception and secretion; it is not difficult to develop a similar theory for the pineal organ of the fishes.

SYNOPSIS

The pineal body of fishes should be reexamined in the light of recent knowledge concerning the physiology of the pineal in higher vertebrates. The presence of photosensory cells with central connections, the extensive vascularity, high metabolic activity, evidence for both external and internal secretory activity, the pineal involvement in phototaxis, and the diverse effects of pinealectomy on the pigmentary response and the endocrine system together with the evidence for an active tryptophan metabolism favour the view that the fish pineal body, far from being an evolutionary vestige, is an important functional organ. While its small size and simplicity do not indicate it is a visual receptor, it is sufficiently sensory in appearance to suggest a role as a dosimeter of variations in incident radiation. The pineal's location and its sensory cum secretory activity suggest that it is well adapted to play a part in the mediation of light influences on the fish pituitary gland; furthermore, it could probably control the pressure and/or composition of the cerebrospinal fluid. As pointed out by Roth (1964), the phylogenetic trend toward a more glandular structure in the higher vertebrates may reflect only a change in the way the information reaches the pineal rather than a completely unrelated change in function.

2

PITUITARY GLAND

In all vertebrates, the pituitary gland or hypophysis consists of two parts, separable on the bases of embryology, structure, and function. These are the neurohypophysis, a downgrowth from the floor of the diencephalon, and the adenohypophysis originating as an ectodermal upgrowth (Rathke's pouch) from the roof of the embryonic buccal cavity. The two parts meet and enclose between them a mesodermal rudiment which gives rise to their intrinsic blood vessels. Thus the gland is a composite organ, and it has many different endocrine functions.

The adenohypophysis is the site of synthesis, storage, and release into the blood of several different peptide hormones; and the greater part of pituitary histophysiology is concerned with the allocation of each of these hormones to the type of pituitary cell that secretes it. The adenohypophysis is divided into the pars distalis, site of secretion of most adenohypophysial hormones, and the pars intermedia. The neurohypophysis in fishes is rather simpler than in land vertebrates and consists essentially of a hypophysial stalk, suspending the gland from the ventral region of the diencephalon (hypothalamus) and containing an extension of the third ventricle (infundibular recess), and at the distal end of the stalk an enlargement, the neurohypophysial lobe or core, which forms the middle of the gland. The stalk contains the axonal fibers of neurosecretory cells, the cell bodies being located in the hypothalamus. The neurohypophysial core consists largely of the endings of these fibers interspersed with cells termed "pituicytes." The neurohypophysis seems to be in general a storage-release center for materials which are actually synthesized in the hypothalamus and then transported to the neurohypophysial core along the neurosecretory axons. In many fishes, the neurohypophysial stalk is virtually absent, the pituitary then being pressed close to the ventral surface of the hypothalamus, while in a few teleosts (e.g., *Lophius*) the neurohypophysial stalk is extremely long.

Adenohypophysial Histophysiology and Cytophysiology

Purves (1966) has usefully divided the cytological criteria used in the study of the adenohypophysis into two groups. The first category consists of features that are indicators of the specific nature of the functions of individual cell types such as granule size, staining reactions and chemical nature, cell morphology, and reactions to specific physiological alterations; these features are the data of special cytology, which is particularly concerned with allocation of function to each cell type. The second category includes those features which are indicators of the functional state of the cell, indicating high or low rates of metabolic or secretory activity such as nuclear size, nucleolar size, amount of cytoplasmic RNA, state of the Golgi apparatus, and accumulation or loss of secretory granules. These features constitute the field of general cytology. In the study of fishes, as in other vertebrate groups, workers on the pituitary have been concerned with both kinds of criteria. However, more than in the highly worked field of mammalian pituitary histophysiology, most investigations on fishes are still primarily concerned with special cytology. This implies that there is, as yet, no complete general agreement about the functions of the various types of cell distinguishable in the fish pituitary; although perhaps with the greater technical standardizations that have come about in recent years, especially the use of methods developed in mammalian studies by Herlant (1956, 1960), and first applied to the fish pituitary by Olivereau and Herlant (1954, 1960), the actual morphological and tinctorial characteristics of the teleostean adenohypophysial cell types would now be agreed upon by most workers. However, there is still disagreement about the functions of these cell types since the experimental allocation of function to the cells has been attempted systematically in only a few species. It follows that not all workers in this field would agree on a generally applicable functional nomenclature of cell types. In this review, the nomenclature used is the one established partly on functional and partly on tinctorial grounds following experimental studies in the eel, *Anguilla*, and the molly, *Poecilia*. It is a mixed nomenclature, in Purves' terms (1966), based partly on similarities in staining properties of the secretory granules to those in mammalian cells of demonstrated function but mainly (primarily) on the characteristic responses of the cells in fishes to experimental situations designed to alter the secretion rates of the different adenohypophysial hormones. Logically, one is on equivocal ground in applying this nomenclature to species in which such experimentation has not been performed. But in the teleosts, which include the vast majority of fishes, the distribution of cell types within the adenohypophysis is extremely regular so that to a far greater extent than in other vertebrate groups the topographical location of a cell type can support its identification on tinctorial grounds. Nevertheless, it must be recognized that the extension of Olivereau's mixed nomenclature to all teleosts would not yet be accepted by all workers on the fish pituitary, and the reader new to this field should bear this in mind. It seems to us that the alternative, to perpetuate yet another purely numerical or Greek letter morphological system, would present further opportunities for the extension to fishes of the nomenclatural confusions that are rife in pituitary studies on higher vertebrates. We think that tinctorial and locational grounds alone are usually sufficiently certain criteria for the extension of the Olivereau nomenclature from those fishes in which it has an experimentally defined functional basis (*Anguilla*, *Poecilia*, and *Mugil* and for certain cell types a few other species) to the vast majority of teleosts in which the desirable experimental backing is lacking. Unfortunately, there are almost no certain grounds for extending the system to nonteleostean groups.

Frequent reference will be made to the staining properties of the cells. By this is always meant the staining properties of the specific secretory granules elaborated and stored within the cytoplasm. The background colourations of the cytoplasm, or of other inclusions such as lysosomes, are not relevant to the tinctorial classification of pituitary cell types. Since it is obvious and readily demonstrable that the method of fixation can greatly influence the reactions of every type of cell to standard staining techniques, and since the affinity of the secretory granules for the various dyes in common use will be affected by the staining procedure applied, wherever possible the cell types will be described from material fixed in sublimated Bouin-Hollande and stained by the various procedures introduced by Herlant, in particular Herlant's Alizarin blue tetrachrome (Aliz B), periodic acid-Schiff with orange G (PAS-OG), and the combination oxidation-Alcian blue-PAS-orange G (Ox-AB-PAS-OG). The central role played by the PAS procedure in pituitary studies has often been emphasized: In essence, this technique distinguishes cells with glycoprotein-containing granules (PAS positive) from cells with nonglycoprotein granulation (PAS negative), a distinction approximately corresponding to the older division between basophils and acidophils and often marked by the terms "mucoid cells" (PAS +ve) and "serous cells" (PAS –ve). The importance of this distinction for pituitary studies stems from the chemical information we have about the various hormones extracted from mammalian pituitary glands, some of which are glycoproteins [gonadotropins, follicle-stimulating hormone (FSH), luteinizing hormone (LH), and thyrotropin or thyroid-stimulating hormone (TSH)] while others are peptides or proteins with no carbohydrate moiety [prolactin, somatotropin (STH) or growth hormone (GH); corticotropin, adenocorticotropin or adrenocorticotropic hormone (ACTH); and intermedin or melanophore-stimulating hormone (MSH)]. Another important technique is aldehyde fuchsin (AF), usually preceded by oxidation, which has a less secure histochemical basis.

Pituitary Gland in Teleosts

General Organization

The organization of the teleostean gland has been the subject of many reviews, and we shall concern ourselves mainly with the more recent information. The anatomy of the gland at first sight appears almost as varied as the teleosts themselves, but closer examination shows that this superficial variety can be reduced to a common anatomical and histological pattern. A species in which the histophysiology of the pituitary has been particularly fully investigated is the European eel, *Anguilla anguilla*, in a long series of fundamentally important studies by Olivereau. The gland has also been investigated in the cyprinodont *Poecilia latipinna* and *P. formosa* by the senior author and his collaborators. A description of the gland in these two teleosts will serve as a basis for a general account of the teleostean pituitary, and notable departures from these basic types will then be considered.

Eel, Anguilla anguilla

Olivereau (1967a) has recently summarized her extensive experimental investigations of the eel pituitary. "The two primary divisions of the gland are obvious, with the central neurohypophysis interdigitating with the shell-like adenohypophysis. The adenohypophysis may be further divided into two main parts: anteriorly the pars distalis, subdivided into rostral and proximal pars distalis, on the basis of cell types; and posteriorly a pars intermedia. As usual in teleosts, the interdigitations of the neurohypophysis with the pars intermedia are

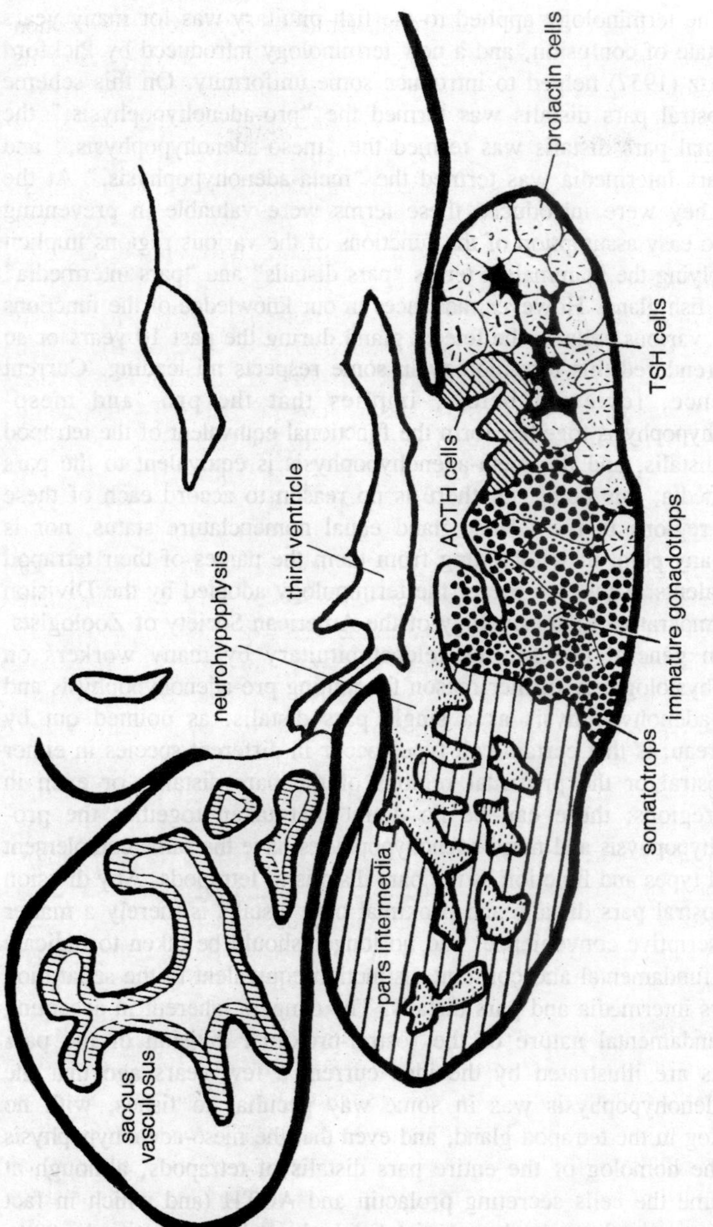

Fig. 2.1. Anguilla anguilla. Diagram of midsagital section through the eel pituitary, anterior to the right.

particularly deep and elaborate, while the neurohypophysial-pars distalis interdigitations are less pronounced.

The terminology applied to the fish pituitary was for many years in a state of confusion, and a new terminology introduced by Pickford and Atz (1957) helped to introduce some uniformity. On this scheme the rostral pars distalis was termed the "pro-adenohypophysis," the proximal pars distalis was termed the "meso-adenohypophysis," and the pars intermedia was termed the "meta-adenohypophysis." At the time they were introduced these terms were valuable in preventing the too easy assumption of the functions of the various regions implicit in applying the mammalian terms "pars distalis" and "pars intermedia" to the fish gland. However, advances in our knowledge of the functions of the various cells in the teleost gland during the past 10 years or so have rendered this terminology in some respects misleading. Current evidence, reviewed below, implies that the pro- and meso-adenohypophysis together form the functional equivalent of the tetrapod pars distalis, and the meta-adenohypophysis is equivalent to the pars intermedia; this being so, there is no reason to accord each of these three regions in the teleost gland equal nomenclature status, nor is there any point in withholding from them the names of their tetrapod equivalents. Thus, we prefer the terminology adopted by the Division of Comparative Endocrinology of the American Society of Zoologists, and in general use for the teleost pituitary by many workers on histophysiology. A further reason for uniting pro-adenohypophysis and meso-adenohypophysis as a single pars distalis, as pointed out by Olivereau, is that certain cell types occur in different species in either the rostral or the proximal regions of the pars distalis, or even in both regions; there can be no doubt that taken together the pro-adenohypophysis and meso-adenohypophyses have the same complement of cell types and function as the pars distalis of tetrapods. Any division into rostral pars distalis and proximal pars distalis is merely a matter of descriptive convenience, and no longer should be taken to indicate some fundamental and constant distinction equivalent to the separation of pars intermedia and pars distalis. The dangers inherent in accepting the fundamental nature of the rostral-proximal division of the pars distalis are illustrated by the idea current a few years ago that the pro-adenohypophysis was in some way peculiar to fishes, with no homolog in the tetrapod gland, and even that the meso-adenohypophysis was the homolog of the entire pars distalis of tetrapods, although at that time the cells secreting prolactin and ACTH (and which in fact both occur in the rostral pars distalis) had not been identified.

In the eel the rostral pars distalis includes three cell types. The η cells (prolactin cells) are arranged in follicles, a feature typical of the more primitive teleosts. Centrally, between these follicles, occur masses and cords of 5 cells (TSH cells), while at the posterior border of the rostral region with the neurohypophysis is a layer of ε cells (ACTH cells). The proximal pars distalis in the eel includes two cell types arranged in cell cords running more or less vertically. Most of the cells are a cells (growth hormone cells), and they are mixed with gonadotrops, of which there are two types. The pars intermedia includes two cell types, the precise functions of which are uncertain. The neurohypophysial fibers contain typical masses and grains of neurosecretory material; even when, as Herring bodies, these accumulations are very large, they are in fact contained within the swollen nerve fibers. This stainable neurosecretory material is usually particularly abundant in the posterior part of the neurohypophysis, where it interdigitates with the pars intermedia. The neurohypophysis also contains scattered cells (pituicytes) of uncertain function.

Molly, Poecilia latipinna and P. formosa

The pituitary gland of this small viviparous cyprinodont is being studied experimentally, and a general account of the gland has been published. The molly pituitary resembles that of the eel, although it is proportionately shorter and deeper. The rostral pars distalis contains only two cell types, a mass of η (prolactin) cells, showing no trace of the follicular arrangement seen in the eel, and a posterior border of ε (ACTH) cells. In contrast to the eel, the δ (TSH) cells occur in the proximal pars distalis, in a dorsal zone where they lie intimately mixed with a cells. Below is a ventral zone of gonadotrops. The pars intermedia is not penetrated by the neurohypophysis so deeply as in the eel and contains two cell types corresponding to those of the eel.

Other teleosts

Departures from the general morphology and cell distribution seen in the above-mentioned two basic species occur, the most obvious being the wide variations in shape and proportions of the gland. A common type has a much shorter antero-posterior axis, and is much deeper dorsoventrally (e.g., cyprinids and salmonids). In such cases, the gland in the juvenile fish may resemble in general that of the eel or *Poecilia*, its shape and proportions changing as the fish grows larger (*Salmo*). In these deeper glands, the neurohypophysial core commonly forms an elongated central axis with the three regions of the adenohypophysis arranged around this axis, the pars intermedia ventral below the hoop of the proximal pars distalis with the rostral pars distalis embracing the neurohypophysial core anteriorly. In these cases, the extensive ramification of the ventral neurohypophysial core into the pars intermedia is very elaborate.

One of the most important variables in the teleostean gland is the location of the δ (TSH) cells, which may lie in the proximal pars distalis, as in *Poecilia* (e.g., *Astyanax*, *Caecobarbus*, *Phoxinus*, and cyprinodonts), or may be rostrally placed, as in *Anguilla* (e.g., clupeoids and cyprinids); or the cells may lie in an intermediate position between the two regions (e.g., cichlids, salmonids, and *Mugil*), this being one of the reasons for uniting the rostral and proximal regions as a single pars distalis. Again, the position of the gonadotrops is variable, being usually in the proximal pars distalis as in *Poecilia* and the eel, but in the trout found in both rostral (few) and proximal regions; and some gonadotrops invade the rostral region even in the eel at full sexual maturity. Another variation concerns the general structure of the rostral pars distalis. In the more primitive teleosts (isospondylous forms, salmonids, clupeoids, and apodes), the prolactin cells of this region are arranged around follicles, as in the eel. This follicular arrangement of the prolactin cells in these primitive teleosts recalls the similar structure of this region in ganoid fishes. In certain clupeoid and salmonid fry, and in the adult *Hilsa ilisha*, the lumina of the rostral follicles communicate with a persistent orohypophysial duct which probably represents the cavity of the embryonic Rathke's pouch. In most teleosts the orohypophysial duct is never present; or if formed it disappears in the adult, and the rostral follicles in the pars distalis of adult isospondylous teleosts should probably be regarded as a primitive feature, as suggested long ago by earlier workers on the gland.

Despite these and other variations in the morphology of the gland, the teleost pituitary when studied in detail usually presents the principal parts described for the eel and *Poecilia*, and the pars distalis can generally be seen to present rostral and proximal regions. In certain cases the distinctions may be less clear, and in *Lepidogobius lepidus* Kobayashi et al. (1959) have described four distinct tinctorial zones in the pars distalis. Probably the experimental analysis of this species would reveal a functional agreement with the more usual roughly bipartite pars distalis.

Lepidogobius demonstrates further unusual features: the whole pituitary is pressed up into the hypothalamus, almost obliterating the third ventricle, the neurosecretory fibers from the hypothalamus take an unusual course to the neurohypophysis, and the neurohypophysial core does not penetrate deeply into the adenohypophysis, not even into the pars intermedia. Another specialized arrangement is seen in *Hippocampus* where the neurohypophysis extends to enclose the pars intermedia laterally, ventrally, dorsally, and posteriorly, so that the posterior region of the gland consists of a central core of pars intermedia enclosed in a sleeve of neurohypophysial tissue.

Histophysiology of the Adenohypophysis

The teleostean adenohypophysis has been shown to secrete the usual complement of hormones, prolactin, growth hormone, gonadotropins, TSH, ACTH, and MSH. Each of these factors has been allocated by experimentation to the cell type that secretes it, and the cell types will now be treated in turn.

Prolactin cells (η cells, erythrosinophilic cells, or paralactin cells)

The prolactin cells lie mainly in the rostral region of the gland, where they form a compact mass that is the main defining character of the rostral pars distalis. In some species, perhaps in most, the prolactin cells extend ventrally and laterally for a greater or lesser distance around the proximal pars distalis, in winglike projections. Their specific secretory granules (η granules) stain red with the erythrosin in Aliz B tetrachrome and the Cleveland Wolfe trichrome, and with the azocarmine or acid fuchsin in other methods such as Azan, Mallory, and Masson, but they are negative to PAS, Alcian blue, and aldehyde fuchsin. In fresh glands of *Poecilia* and *Fundulus kansae* the η granules are opaque so that the rostral part of the pituitary is often a dense white. In the isopondylous forms in which the prolactin cells are arranged in follicles, the material commonly found in the follicular lumen exhibits variable staining, some with orange C and some with PAS, Aniline blue, or light green; with the electron microscope, at least some of this material in the eel consists of a mass of membrane-bound vesicles, probably derived from degenerating cells. The prolactin cells in such fish (e.g., eel, trout, and salmon) are generally columnar, their apices bearing cilia projecting into the follicular lumen, and often with the η granules concentrated toward the outer cell base. The cytoplasmic RNA (i.e., endoplasmic reticulum) usually lies between the nucleus and the outer base of the cell. Ultrastructural studies on the eel revealed electron dense granules in these cells, about 280 mμ diam in freshwater *Anguilla* but about 350 mμ in the marine *Conger*; the endoplasmic reticulum lies basally, while the Golgi apparatus lies on the other side of the nucleus, toward the follicular lumen. The Golgi apparatus in immature freshwater eels consists of six to eight parallel cisternae forming a bowl- or cup-shaped body; ultrastructural appearances suggest that the η granules are formed here, the outer convex surface of the Golgi perhaps receiving newly synthesized material from the closely adjacent rough endoplasmic reticulum, the concentrated materials then being released as membrane-bound granules at the inner concave surface of the Golgi. In *Anguilla* elvers, and in adult *Anguilla* caught in the sea during the spawning migration, the η granules are smaller than in the freshwater stage, about 200 mμ diam.

The prolactin cells of nonisospondylous teleosts are not grouped in follicles and are not generally columnar (although the cells may be elongated in some species such as the cichlid, *Heterichthys cyanognathus*). Usually they form a compact mass of rounded cells as in *Poecilia*, generally evenly granulated. In cyprinodonts, the nucleus is frequently indented or kidney-shaped, a feature particularly marked in *Poecilia* and *Fundulus heteroclitus*; in these cyprinodonts, too, the Golgi image is easily visible in active cells as one or more clear tubules among the granules, curved in a C or U shape, and the endoplasmic reticulum, visualized by staining for RNA, forms a cap or halo on the nucleus. In the related guppy (*Poecilia* = *Lebistes reticulatus*) and platy (*Xiphophorus maculatus*), the η granules are revealed by the electron microscope as membrane-bound osmophilic vesicles, between 200 and 300 mμ diam, and Weiss (1965) has described ultrastructural appearances that suggest that the η granules in the platy are probably released from buds projecting from the cell surface, by a process in which the granule membrane fuses with the cell membrane and the granule escapes through the resulting opening; this resembles one of the modes of granule extrusion described in higher vertebrates.

At the light microscope level, η granules appear to vary in size even within individual cells, and commonly vary between individuals and species; in *Poecilia latipinna*, when the granules are rather sparse they appear to be distinct and large, but if they are densely packed they seem smaller, while in the eel and trout the granules are usually coarse and distinct no matter what their density. Here, and in other places in this review, we should emphasize that granule size, even if determined with the electron microscope, is not an exact criterion of cell type in comparisons between species. Fluorescent antibody to ovine prolactin located specifically on the η granules in *Fundulus heteroclitus*, confirming that the granules themselves do contain fish prolactin.

Histochemically the prolactin cells have been shown to contain SH/SS groups in the eel, Mugil, and *Xiphophorus*; but as in other species they do not react with Alcian blue even after oxidation, indicating that they are not particularly rich in cysteine. Nevertheless, *Mugil* prolactin cells will incorporate cysteine-^{35}S. In goldfish in freshwater these cells displayed intense incorporation of acetate-^{3}H, indicating intense protein synthesis. The η granules in fresh *Mugil* glands were precipitated only by 7.5% or stronger trichloroacetic acid, in contrast to the α granules (growth hormone cells) which were precipitated by 2.5% TCA. In ultrastructural studies of eel prolactin cells, acid phosphatase was shown to be distributed in particles in the apical cytoplasm, and also within the Golgi region and in some of the developing η granules, possibly concerned in the destruction of excess secretory material as suggested for the lysosomelike bodies in the gonadotrops.

Evidence for the secretion of fish prolactin by the η cells

As early as 1960, Olivereau and Herlant pointed to the similarity in staining properties of the teleostean η cell and the prolactin cell of mammals. However, too little was known at that time about the physiological role of prolactin in fishes to suggest an experimental approach to defining the function of the η cells.

More recently, following the demonstrations by Burden (1956), Pickford and Phillips (1959), and Pickford et al. (1965) that prolactin is the only mammalian pituitary hormone that will promote tolerance of freshwater in hypophysectomized *Fundulus heteroclitus*, the η cells have been investigated experimentally, especially in *Poecilia latipinna*. Details of the physiological background will be found in the chapter by Ball, this volume. For the present, the important point is that evidence indicates the secretion by the pituitary in both *F. heteroclitus* and *P. latipinna* of a prolactinlike hormone (fish prolactin, paralactin) that specifically promotes survival in freshwater by limiting the outflux of sodium from the body. Fish prolactin is not essential to *Fundulus* in seawater nor to *Poecilia* in dilute seawater, but it is secreted, presumably at a low rate, by *Poecilia* in dilute seawater, with the effect of reducing the rate of sodium exchanges.

Studies on the pituitary in these two fishes have shown that only the η cell displays cytological evidence of changes in secretory activity such as would be predicted of the cells secreting fish prolactin. Thus, in *Poecilia* these cells are always more active in freshwater than in dilute or full-strength seawater; and these cells, but no others, are rapidly activated when *Poecilia* enters freshwater from dilute seawater, in correlation with reversal of plasma sodium loss, a marked curtailment of sodium outflux from the body, and an increase in pituitary prolactin content. Similarly, in *F. heteroclitus* the η cells are consistently more active and numerous in freshwater than in seawater, and a regenerated pituitary remnant in an incompletely hypophysectomized individual, experimentally shown to secrete fish prolactin, consisted almost entirely of active η cells. In more direct experimental approaches, it was shown that removal of part of the zone of η cells impaired freshwater tolerance of *P. latipinna*, and that ectopic pituitary transplants in *P. formosa* and *P. latipinna* are able to secrete fish prolactin, the transplants always containing active η cells. The identification was clinched when Ball (1965c) demonstrated that ectopic pituitary transplants of the rostral part of *P. latipinna* pituitary, composed mainly of η cells, secreted fish prolactin in response to entering freshwater, but that transplants of the posterior part of the gland, containing few η cells, did not. More recently, Emmart et al. (1966) showed that fluorescent rabbit antiserum to ovine prolactin located only on the η granules in *F. heteroclitus* and in no other cell type. Thus, the functional identity of the η cells in these cyprinodonts is as well established as that of any pituitary cells in any vertebrate, and there can be no doubt that the η cells secrete fish prolactin.

In other teleosts, these cells behave in the same way as in the cyprinodonts. They are more active in freshwater than in seawater in salmonids and in the eel, in *Mugil*, in *Fundulus kansae*, and in *Tilapia mossambica*. They occur in all teleosts that have been examined, as far as can be judged from the published description sometimes based on material not suitably fixed and stained. In some species, usually marine, the prolactin cells may appear chromophobic, especially after poor fixation; and when the eel enters seawater the η granules first become smaller and then virtually disappear, leaving apparently inactive chromophobic cells. Conversely, the granulation is sparse in elvers newly arrived in freshwater from the sea and the cells may appear totally chromophobic in young eels, but in the freshwater phase of larger eels they are always well granulated.

There are a few observations on changes in the η cells in relation to life history. They are strongly granulated and erythrosinophilic in newly born *Poecilia latipinna* and are already differentiated in the gland of the embryo within the maternal ovarian follicle; they differentiate early in gestation in the guppy embryo (*Poecilia reticulata*) at first with granules smaller than in the adult, and with an active-type endoplasmic reticulum. The prolactin cells do not appear to undergo marked activity changes during the monthly cycle of oocyte growth and pregnancy in viviparous cyprinodonts (*Xiphophorus*); but in *Zoarces viviparus*, an unrelated viviparous form with a totally different mode of gestation, ultrastructural studies indicated that these cells are hyperactive during pregnancy. In the male *Hippocampus*, which incubates the eggs in a brood pouch or marsupium, the prolactin cells undergo an annual cycle in correlation with the development and functions of the marsupium, being especially active during the first half of the incubation period. These changes correlate with experimental evidence that prolactin is concerned in maintaining the marsupium. The η cells of salmon, *Salmo solar*, are fairly active in the freshwater parr and smolt and appeared to be reorganizing after degeneration in adult fish ascending the river from the sea in the spring, eventually becoming stimulated and hyperplastic. Their activity seemed to be depressed in spawning fish in freshwater, particularly in the female. The η cells in hybrid *Xiphophorus* bearing melanomas were extremely large and hyperactive, an interesting correlation in view of evidence that prolactin can promote melanogenesis in *Fundulus heteroclitus*.

In view of all the evidence for the part played by fish prolactin in sodium conservation in freshwater, one might expect the η cells to be maximally active in deionized water; but this is not so, at least in the case of the eel, in which a sojourn in deionized water reduced the cells to a state of inactivity comparable to that in seawater, with marked concomitant changes in other cell types described elsewhere. Deionized water is obviously a highly artificial medium, and in its effects on electrolyte metabolism it does not act simply as a highly dilute freshwater.

There are indications of interactions between the prolactin cells and other endocrine glands. The cells are activated after radiothyroidectomy in the eel and goldfish, but not distinctly so in the trout. However, thyroxine had no very clear effect on the cells in the eel apart from inducing a retention of η granules. Doses of thiourea or thyroxine, both of which inhibited thyroidal ^{131}I uptake, had no obvious effects on the prolactin cells in *P. latipinna*; nor did propyl thiouracil affect the prolactin cells of *Mugil*, although inducing pronounced alterations in the TSH-thyroid axis. Surgical interrenalectomy did not alter the prolactin cells in the eel, although the adrenal inhibitor SU 4885 leads to slight and irregular stimulation of these cells. Given over three days, ACTH inactivated the prolactin cells of *Hippocampus*. The only indications of gonadal influences on these cells are Olivereau's observation (1967a) that in freshwater the prolactin cells are more active in female than in the male eels (immature fish), and Sokol's report (1961) that the η cells of *F. heteroclitus* undergo a transient chromophobia (= partial degranulation) coincident with spawning in the coastal seawater.

Like the mammalian prolactin cell, the η cell in *Poecilia* remains active in ectopic pituitary transplants, an activity manifested both functionally and cytologically. In such transplants, as in the normal gland, the η cells are activated rapidly when the fish enters freshwater from dilute seawater. Thus hypothalamic connections are not essential for activation of the η cells in response to salinity reduction, nor for the maintenance of their activity in freshwater. However, we are not in a position to postulate a hypothalamic prolactin-inhibiting factor (PIF) such as exists in mammals, since we do not yet have quantitative data on secretory rates of fish prolactin from transplants and from the *in situ* gland. It is difficult, because of this, to interpret Olivereau's observations that prolonged treatment with ovine prolactin leads to regression of the η cells of *Anguilla*, and Boisseau's finding (1967) that a three-day treatment with prolactin led to marked involution of *Hippocampus* η cells. Comparable physiological observations in the rat are suggestive of an increase in secretion of PIF induced by the exogenous prolactin, leading to inhibition of endogenous prolactin output, but obviously this concept may not be extended to teleosts in the present state of our knowledge.

In vitro observations agree with the results from pituitary transplants, the η cells remaining active in cultured trout glands, with partial or total degranulation and cellular enlargement. In cultured glands from *Fundulus heteroclitus*, Emmart and Mossakowski (1967) found that new colonies of η cells, arising by mitoses in the outgrowing layers of the explant, contained granules that bound a fluorescent antibody to sheep prolactin, indicating the continued ability of the η cells to synthesize fish prolactin in the complete absence of hypothalamic influences. Fixed material from these cultures showed evidence of release of fluorescent-labeled granules across the cell surfaces, and in addition these workers observed with phase contrast the emission of granules from the constantly rippling and undulating membrane of the living cell. Sage (1968), working with *Xiphophorus* glands, used elaborate techniques to show that the η cells lost material *in vitro*, and that this loss was greater on more dilute media than on a medium with approximately the same sodium content as the fish serum. He did not specifically determine whether dilution of the medium increased the synthetic rate of the cells, in addition to enhancing their rate of discharge (although this is implied by their degranulation in the presence of cyanide); certainly, in *P. latipinna* transferred to freshwater from dilute seawater, plasma sodium falls through the range of concentrations employed by Sage concomitantly with the onset of increased discharge and synthesis in the prolactin cells.

ACTH cells (Corticotrops, ε cells, or X cells)

The ACTH cells, characteristically disposed in a sheet at the interface between the rostral pars distalis and the neurohypophysis, have been recognized morphologically in teleosts for many years, but their significance was not known. Some workers thought that they were aberrant or immature forms of the η cells, while Bugnon suggested they might represent the pars tuberalis of higher vertebrates. In recognition of their enigmatic nature, Olivereau (1963b) termed them the "X cells."

In *Poecilia*, the ε cells are usually truly chromophobic (that is, without granules that can be stained by any of the standard dyes or histochemical techniques); but sometimes they display a very fine, powderlike granulation, faintly staining with erythrosin in Cleveland Wolfe or Aliz B preparations, or with azocarmine in the Azan technique. They may also stain a faint purple after Aliz B or gray with lead hematoxylin (PbH). They are elongated cells, usually arranged with the long axis perpendicular to the adeno-neurohypophysial interface. The cells are small, and the absence of distinct granulation makes it difficult to discern the cell outlines. The most characteristic feature in this species is the elongated nucleus with a distinctive scattering of chromatin which gives the nucleus an easily recognizable spotted appearance. The nucleolus in most normal cells is insignificant or undetectable.

In the eel the ε cells are much better granulated than in *Poecilia*, and the granulation stains weakly with erythrosin and quite strongly with Alizarin blue. The granulation is also strongly stained by PbH which picks out the cells easily, although it is not a specific stain for the ACTH cells; PbH also stains a gonadotrop in the eel and a pars intermedia cell type in most teleosts. The eel cells are usually oval rather than fusiform as in *Poecilia*. The nucleus may be rounded, elliptical, or pear-shaped, and the nucleolus is generally indistinct. In all teleosts in which they have been described, these cells display no trace of an affinity for PAS or Alcian blue; that is, they are nonmucoid or serous cells. In *Anguilla*, they are rich in SH/SS groups.

ε Cells were described in *Xiphophorus* as polymorphic chromophobes, forming a wedge-shaped area posterior to the η cells; a similar arrangement was found in the surfperch, *Embiotoca*. In the Pacific salmon they border the neurohypophysis behind the mass of η cells, and are rather insensitive to dyes, with a fine granular cytoplasm that stains brown after Masson. Baker (1963a) described trout ε cells as forming a layer of amphiphils interspersed between η cells and neurohypophysis and coloured by Alizarin blue and PbH. A similar account of the salmonid ε cells was given by Olivereau (1964a), who emphasized that their affinity for Alizarin blue and PbH is much less than in the eel and also that they contain fewer SH groups than in the eel. Typical ε cells have also been described in *Perca* and in *Mugil*, and in the goldfish they are small and apparently chromophobic elements among the η cells, but they will stain with Alizarin blue and PbH. Oztan (1966a) described ε cells in *Zoarces* as columnar acidophils, facing the rostral pars distalis and staining with light green; however, she was not able to separate them from the η cells with the electron microscope. *Fundulus heteroclitus* also has typical ε cells, in a palisade behind the η cells. Sokol (1961) described them as blue-green basophils; (after aldehyde fuchsin-light green-orange G); and they also display the characteristic affinities for erythrosin, Alizarin blue, and PbH. Sokol also described the ε cells of the guppy, *Poecilia reticulata*, as weakly staining green basophils.

At the ultrastructural level, the eel ε cell reveals a moderately electron dense cytoplasm containing numerous fine vesicles or microtubules. The endoplasmic reticulum is diffuse and not prominent, and the Golgi area is rarely seen in normal fish. The cytoplasm is evenly packed with electron dense granules measuring 200-250 mμ in diameter. Follenius (1963a) described the cells as forming a chromophobic band in several teleosts, and his ultrastructural studies revealed a sparse cytoplasm with ergastoplasmic vesicles in some cells; other cells contained a well-developed Golgi apparatus surrounded by osmiophilic secretory granules. The most general point of agreement in the various accounts is that these cells tend toward chromophobia, especially marked in cyprinodonts. This tendency may be exaggerated by stress at sacrifice, when the animal may struggle, or by prolonged anesthesia. A similar tendency toward chromophobia is characteristic of the ACTH cell in higher vertebrates.

Evidence for the secretion of ACTH by the ε cells

Allocation of ACTH secretion to these cells was accomplished simultaneously and independently by treatment of *P. latipinna* and

Anguilla with Metopirone, a drug which inhibits the biosynthesis of adrenocorticosteroids. Although other actions have been described, its main effect in mammals, following short-term treatment with low doses, is inhibition of 11β-hydroxylase, thus preventing the formation of the normal 11-oxycorticosteroids. As a result, because of the negative feedback relationship between these steroids and ACTH, the use of SU 4885 has led to identification of the ACTH cells of mammals and birds.

In *Poecilia* and the eel, short-term treatment with SU 4885 resulted in hypertrophy and hyperplasia of the interrenal cells, indicating increased ACTH secretion, and this was accompanied specifically by activation of only the ε cells in the pituitary. This morphological stimulation of the interrenal by the drug is partly (eel) or wholly (*Poecilia*) suppressed by hypophysectomy, which supports the interpretation that SU 4885 elicits elevated ACTH output and hence identifies the ε cells as corticotrops. With longer treatment in both species the TSH cells are stimulated, along with thyroidal activation, and other effects of the drug have been observed with long treatment in the eel. However, there can be no doubt that the initial and most marked hypophysial effect of SU 4885 is to provoke enhanced secretion of ACTH, and that this originates in the ε cells. Surgical removal of the interrenal in *Anguilla* has confirmed this earlier work, resulting in activation of the ε cells specifically. However, interrenalectomy, unlike SU 4885 treatment, did not lead to stimulation of the TSH cells, suggesting that the action of the drug on these cells may be direct rather than mediated via the ACTH-interrenal axis. Unpublished work on *P. latipinna* has shown that if cortisol treatment is combined with SU 4885, the interrenal stimulation (i.e., elevated ACTH secretion) is prevented, and correspondingly the ε cells are not activated.

This identification has been further strengthened by studies in the eel of the responses of the ε cells to a variety of other treatments designed to alter ACTH-interrenal function; the cells were activated by treatment with reserpine, aldactone, op'DDD, and inactivated by treatment with cortisol or ACTH. Aldactone also activates these cells in goldfish, and a heavy infection of the kidney, which partly destroyed the interrenal, was associated with ε cell activation in a Pacific salmon. In *Hippocampus*, as in the eel, ACTH inactivates the ε cells.

The degranulation of the ACTH cells following conditions of stress such as anesthesia and bleeding was emphasized by Olivereau (1967b), and such a stress response could perhaps explain the slight stimulation of these cells when *Mugil* was transferred from seawater to freshwater. The stress response of the ACTH cells was even more marked in response to surgery in the eel, with correlated activation changes in the interrenal.

A lengthy sojourn in deionized water, which had profound effects on other cell types, produced only slight stimulation of the ε cells and interrenal in *Anguilla*, perhaps indicating that the ACTH-interrenal axis plays only a minor role in electrolyte regulation in this dilute medium.

There are a few observations on the response of these cells to collateral endocrine changes. They are slightly inactivated after radiothyroidectomy in the eel and trout, and thyroxine treatment led to slight stimulation of the eel ACTH cells, with nuclear hypertrophy, slight degranulation, and a few mitotic figures. In *Mugil*, too, the cells were stimulated by thyroxine and inhibited by thiourea. Removal of the enigmatic corpuscles of Stannius from eels led to changes in the ACTH cells that were variable and difficult to interpret. Also in the eel, the ACTH cells became more active with sexual maturation induced by injections of carp pituitary material, but interpretation of this effect is complicated since the carp material probably included ACTH, indicated by the fact that the interrenal tissue was more strongly stimulated than the degree of activation of the ACTH cells would explain. Boisseau (1967) described an annual cycle of activity in the ACTH cells of the male *Hippocampus*, related to the annual sexual cycle and characterized by hyperactivity of the ACTH cells associated with incubation of eggs in the marsupium. Boisseau (1967) also found that prolactin injections led to inactivation of the ACTH cells. This detailed work on *Hippocampus* is exceptionally interesting in indicating the way in which the ordinary complement of pituitary hormones may be adapted to control extremely specialized mechanisms: The maintenance of the marsupium (developed under testicular influence, in turn dependent on gonadotropins) is dependent upon the ACTH-interrenal axis and upon prolactin. Boisseau found evidence of complex interrelationships between prolactin, ACTH, and corticosteroids in *Hippocampus*; and the response of the η and ε cells to ACTH and prolactin administration may in themselves be specialized in connection with the highly unusual reproductive processes. Thus, prolactin treatment had no obvious effects on the ACTH cells in the eel, unlike *Hippocampus*.

The ACTH cells persisted, although in reduced numbers, in ectopic pituitary transplants in *Poecilia formosa* which continued to secrete small amounts of ACTH. However, these cells in the eel pituitary *in vitro* present an appearance of *increased* activity, showing degranulation, increased nuclear size, and rapid incorporation of radiouridine. In the trout the ACTH cells in culture display a marked migration of their granules toward the pole of the cells facing the neurohypophysis; this phenomenon is difficult to interpret, but recent unpublished work by Baker shows that, as in the eel, these trout cells displayed intense nuclear incorporation of radiouridine. Nevertheless, the granulation continues to accumulate at the pole of the cell, which suggests inhibition of release, in contrast to the degranulation of the eel cells. Thus there may well be marked species variations in the response of the ACTH cells to hypothalamic disconnection.

Growth hormone cells (somatotrops, α cells, GH cells, or STH cells)

The α cells are the most prominent acidophils in the pars distalis, especially in marine fishes with inactive prolactin cells, and occupy much of the proximal pars distalis. They have been described and recognized as a distinct cell type for many years and in many species. These are the cells that conform to the classic category of acidophil cells, staining intensely and selectively with orange G in the various trichrome and tetrachrome techniques. For many years the existence of two types of acidophils has been recognized in mammals and other tetrapods, one type staining selectively with orange G (α cells) and the other type (η cells) with azocarmine or erythrosin. However, the distinction is not always easy to obtain in all species and with all dyes and techniques; and although the distinction can be made in teleosts, here too it is delicate and not always easy. In the case of well-known pituitaries, such as *Anguilla* or *Poecilia*, this difficulty is not always important, since once established that one can in some circumstances make this tinctorial differentiation, one can always separate the α and η cells on the basis of their location and morphology; but with any new or iittle-known teleostean gland it is advisable to spend time and effort in establishing this point before making definitive identification of the α and η cells. It needs to be emphasized that the pituitaries of different fishes vary greatly in their staining reactions to any particular techniques, and one must experiment with dyes, timings, and other variables for each species.

In *Poecilia* the η-α tinctorial distinction can be made quite easily with the Azan or Aliz B method, the η cells staining red with azocarmine or erythrosin and the α cells staining orange (Azan) or clear yellow (Aliz B). A similar distinction can be obtained in the eel, trout, salmon, *Xiphophorus*, *Fundulus kansae*, and *F. heteroclitus*. Emmart et al. (1966) emphasized the delicacy of the distinction, showing that the red-yellow differentiation obtained with PAS-erythrosin-orange G could be reversed by rinsing the slide initially in 1% boric acid. In *Poecilia* and *F. heteroclitus*, the combination Ox-AB-erythrosin-orange G usually gives a good α-η distinction, as well as staining the basophils m a sharply contrasting blue.

In *Poecilia* the a cells occur in the centro-dorsal region of the proximal pars distalis, intermingled with the much scarcer TSH cells. With the light microscope the cells are generally large and rounded in form, although frequently elongatcd and ellipsoid; typically, the cell is crowded with exceedingly fine orangeophilic granules, finer than the η granules and often packed so densely as to appear a homogeneous mass. The nucleus is ovoid, frequently eccentrically placed at one pole of the cell, and often seems squashed and distorted by the dense packing of the granules. In any one fish there may be great variations in the size and shape of the nucleus, and in the size of the nucleolus; the latter is generally prominent, however, and the cells usually show signs of considerable activity. The Golgi image, in the form of a ring or crescent, is usually very prominent, sharply outlined by the dense granulation.

A tendency to sexual dimorphism is evident in the α cells of *Poecilia*: In the adult female, the cells are numerous, often tightly packed together, densely granulated and very active in appearance, as already described, whereas the male fish usually presents few α cells, poorly granulated or even degranulated, with a central nucleus, nucleolus insignificant or invisible, and no sign of the Golgi image. The α cells mainly lie in the dorsally projecting fingers of the proximal pars distalis, hence in close proximity to the neurohypophysis. In the female, they frequently form a virtually continuous border between the neurohypophysis and proximal pars distalis, several cells in depth, forming a shell, dorsally concave, which is easily visible in the middle of a fresh gland, the a granules being translucent and appearing a dense white in contrast to the more transparent ventral gonadotrops. Similarly in mammals, acidophil granules appear opaque in fresh material. In contrast, in the male *Poecilia* the α cell border to the proximal pars distalis is thinner and discontinuous, the cells are poorly granulated and generally not visible m the fresh gland.

In the sexually mature female eel, as in *Poecilia*, the α cells display a sexual dimorphism. In immature individuals of both sexes, these cells form the greater part of the cords of the proximal pars distalis, since the TSH cells are located in the rostral region, and the gonadotrops are small and undifferentiated. At sexual maturity the cords become composed of a mixture of α cells with now more numerous and larger gonadotrops. In the mature male, the α cells are small, with indistinct boundaries, rather sparse granulation, and indistinct nucleolus and little or no visible cytoplasmic RNA, whereas in the female the α cells are much more numerous and often appear more active, with strongly marked cell boundaries, large nuclei and prominent nucleoli, denser granulation, prominent Golgi image and demonstrable cytoplasmic RNA. Thus in both *Poecilia* and *Anguilla*, the a cells are generally more numerous and more active in the mature female. In contrast, Schreibman (1964) found no indications of such a sexual dimorphism in *Xiphophorus*.

In both *Poecilia* and *Anguilla* the a granules are more refractile than the η granules. In contrast to *Poecilia*, the eel α granules are rather larger than those of the η cells, an observation confirmed by the electron microscope. Similar differences are found in comparisons of the two cell types in other teleosts, some fish resembling *Poecilia* in having smaller α granules, while others resemble the eel in having smaller η granules (*Perca*, cichlids). This variability again emphasizes that granule size alone is not a criterion of cell type.

Like the η cells, the α cells are typical serous cells; that is, their granules do not contain glycoprotein and are typically negative to PAS, AF, and AB. A faint staining with PAS can nevertheless be obtained, especially with the Ox-AB-PAS-OG procedure, in *Poecilia* and *Fundulus*, the eel and other species. This recalls the faint staining of acidophil granules with PAS in mammals, which may reflect the lipid content of these granules. This PAS reaction in the teleostean α cell is too weak to cause confusion with the strong reaction in the mucoid cells. The α cells in *Anguilla* contain SS/SH groups and are rich in protein, and in *Mugil* they incorporate radiocysteine rapidly. Tritiated acetate and labeled amino acids were rapidly accumulated by the α cells of goldfish, evidence of intense protein synthesis. *Mugil* α granules are precipitated by weak (2.5%) trichloroacetic acid, in contrast to the η granules which precipitate only with a stronger (7.5%) solution; and Follenius (1967) found that the α cells were selectively labeled following intraperitoneal injections of D,L-noradrenaline-^{3}H in *Gasterosteus*. The significance of this finding is uncertain at present.

α Cells have been recognized by nearly all workers on the teleost pituitary, and a resume of all their accounts would be unnecessary.

Certain ultrastructural observations are important. In *Anguilla*, Knowles and Vollrath (1966b) observed that the α cells appeared to be filled with spherical or slightly oval membrane-bound vesicles, about 400 mμ in diameter. No clear indications of endoplasmic reticulum or Golgi apparatus could be found, and there were few mitochondria. Follenius (1963a) described ultrastructural variations which he thought might indicate a cycle of activity in the α cells of *Perca*. Cells rich in endoplasmic reticulum were poor in secretory granules, and vice versa; in cells that were primarily storing their secretory products, the endoplasmic reticulum was reduced to a few cysts, often perinuclear in position. Follenius described granules of secretion forming in the Golgi region, and also extrusion of these granules from the cell surface in a process reminiscent of that described by Weiss for the teleostean η cell, Follenius emphasized that the α cells present greater variations in activity than the η cells, but he did not study the responses of the η cells to changes in salinity.

Evidence for the secretion of growth hormone by the α cells

A few years ago, Olivereau (1963a,b) summarized the evidence that the teleostean α cell secretes growth hormone. As she said, there was then no direct proof of this ascription, although the similarity of the α cells to the somatotrops of mammals made it probable. In teleosts, the α cells are more numerous in salmon during the period of rapid growth in the smolt stage and were also very active in *Oncorhynchus* captured at sea during the adult period of rapid growth. The greater abundance and activity of these cells in the female eel may relate to the greater size attained by the female in freshwater, and the generally larger size of the female may similarly explain the sexual dimorphism of the α cells in *Poecilia*.

More recently, with the experimental evidence that has been derived for the functions of the other pars distalis cell types, the α cells have come to be accepted as somatotrops, as it were by a process of elimination. More specifically, when the cellular source of gonadotropin(s), TSH and ACTH were known, the two acidophils, α and η cells, remained as the producers of prolactin and growth hormone; we have seen that the η cells secrete fish prolactin, which leaves the α cells as the source of growth hormone. The functional distinction between the two acidophils was clinched by partial hypophysectomy in *Poecilia latipinna*, which showed that removal of most of the rostral pars distalis (η cells), while impairing prolactin secretion, had no deleterious effect on growth. This meant that growth hormone must be produced in the proximal pars distalis, and knowing the gonadotrops, TSH cells, and ACTH cells, this leaves the α cells as somatotrops. The scarcity and degranulation of α cells in the ectopic pituitary transplants that secreted little growth hormone in *P. formosa* is in agreement with this conclusion.

Unfortunately, little information is available about alterations in the a cells during the life history. The relationship of these cells to periods of rapid growth in salmon has already been mentioned. The α cells of the guppy, *Poecilia reticulata*, differentiate and attain their adult condition later than in the η cells; they are present and active in newly born and rapidly growing *P. latipinna* fry, and they have been described as staining more intensely in the elver than in the adult yellow eel. The α cells were enlarged and particularly active in melanoma-bearing hybrid *Xiphophorus*, a suggestion of a possible tumorogenic action of fish growth hormone.

Concerning the response of the α cells to collateral endocrine changes, Olivereau detected no action of thyroxine or radiothyroidectomy on these cells in the eel, and female *P. latipinna* display no marked a cell reactions to thyroxine or thiourea treatment. Radiothyroidectomy did not affect the α cells in trout. However, Sage (1967), while agreeing that thyroxine alone had no action on the α cells of male guppies, *Poecilia reticulata*, found that thiourea caused increased storage of granules in the cells, indicating retention of growth hormone, and that thyroxine given in combination with thiourea reduced this storage of granules. Since thyroxine increased the growth of thiourea-treated guppies. Sage interpreted his findings as evidence that thyroxine in some way is necessary for the release of growth hormone from the α cells. However, on this hypothesis and with this type of evidence, the enhanced growth produced by thyroxine alone (and repeatedly observed in thyroxine-treated female *P. latipinna* also) cannot be attributed to increased release of growth hormone, since the α cells in the guppy, as in *P. latipinna*, showed no response to thyroxine by itself. Most probably thyroxine synergizes peripherally with normal levels of growth hormone to cause growth enhancement, in the same way that exogenous bovine TSH and hake GH synergize in hypophysectomized *Fundulus heteroclitus*.

Treatment with cortisol causes regression of the α cells in the eel, while surgical stress activates them, although not the stress of anesthesia and bleeding. Surgical removal of the eel interrenal did not appear to activate the α cells any more than the ham operation, although the adrenocortical inhibitor SU 4885 slightly increased the α granulation. At sexual maturity the female eel α cells are reduced in number, which may correspond to a reduced growth such as commonly accompanies sexual maturation. Sokol (1961) described a marked cycle in the α cells of the female guppy in relation to gestation, the cells being mainly inactive during the greater part of the month-long gestation period, but showing transient activity at the time of parturition, while in the unrelated viviparous *Zoarces* Oztan (1966a) described marked activation of these cells during gestation and regression following parturition.

Eels with small skin lesions exhibited hyponatremia accompanied by strong activation of the α cells, while, surprisingly, extreme activation of these cells was found in eels maintained in deionized water for up to 46 days. Prolonged exposure to deionized water leads to marked ionoregulatory disturbances in the eel, apart from hyponatremia, including a fall in plasma calcium with a rise in plasma potassium levels that seems to result from movement of potassium from muscles to plasma. Olivereau suggested that the increased secretion of growth hormone in deionized water might represent a response designed to oppose this internal shift of potassium, since the rise in plasma potassium occurs more rapidly in hypophysectomized eels than in intact animals, but for the moment the real significance of this reaction of the eel α cells is uncertain, and its explanation awaits a full-scale investigation of the possible role of growth hormone in teleostean electrolyte metabolism.

The α cells were reduced in number and relatively inactive in pituitary transplants in *Poecilia formosa*, and in correlation these cells frequently regressed in the trout pituitary cultured *in vitro*.

Gonadotrops (β and γ cells)

When adequately identified, the gonadotrops of teleosts have usually been found located in the proximal pars distalis, and most often in the ventral part of this region. However, gonadotrops also spread into the rostral pars distalis at sexual maturity in the eel, salmon, and trout. They are typical mucoid cells containing granules that react strongly in the PAS procedure and are assumed to contain glycoproteins. For most practical purposes, this property amounts to a redefinition of the old category "basophil cells," just as the absence of typical PAS +ve granules defines the serous cells which are equivalent to the old category of "acidophil cells." With this definition in mind it is often convenient to continue using the terms "acidophil" and "basophil," as long as one remembers that the terms have no reference to affinities for acidic or basic dyes; these older terms are so deeply entrenched in general usage that their replacement by "serous" and "mucoid" cells, although frequently advocated, is unlikely to materialize. In nearly all cases reported, gonadotrop granules in teleosts stain with PAS, AF, and AB, and with Aniline blue in the common trichrome and tetrachrome techniques.

The appearance of the gonadotrops varies greatly during the cyclic development of the gonads, and the cells are best characterized by an account of their annual or shorter cyclic changes. However, one very important point to be dealt with first concerns the question of whether teleosts possess only one type of gonadotrops or two. Higher vertebrates, with the established possession of two gonadotropic hormones (FSH and LH), have generally been found to have two kinds of gonadotrops in the pituitary, the β cell secreting FSH and the γ cell secreting LH. Since there is in fact no very definite physiological or biochemical evidence for the presence of two distinct gonadotropins in teleosts, even in those fishes which have been found to possess two distinct kinds of gonadotrops, one may not apply the terms "beta" and "gamma" to these cells.

The eel and the Pacific salmon are two of the teleost species in which gonadotrops can be separated into two types, with characteristics summarized. It will be seen that in these cases the distinction depends not only on tinctorial features but also on position, cell size, and granule size and morphological features such as vacuolation. The two types in the eel appear to be equivalent in both males and females. Measurements showed that the eel Type 2 cells tend to be smaller than the Type 1 cells, but with larger nuclei. At the electron microscope level, two kinds of gonadotrops have been separated in the eel, with granules of about 130 and 190 mμ diam, probably corresponding to the light microscope categories of Olivereau. Ox-AB-PAS-OG distinguishes two gonadotrops in the goldfish, as in the eel and salmon. Two types are also discernible in *Mugil auratus*: In this fish, one kind of gonadotrop is situated medially in the gland and is coloured a clear blue-green after Aliz B, and also stains with PbH. After spawning this cell becomes vacuolated. The second gonadotrop is placed more laterally in the gland, stains purple with Aliz B, is negative to PbH, and rarely becomes vacuolated. Different techniques applied to another mullet, *M. cephalus* also separated two gonadotrops: with the aldehyde thionine-PAS-naphthol yellow method (AT-PAS), one type lying at the periphery of the cell cords of the proximal pars distalis stains brick red, while the other type lies in the center of the cord and stains dark violet. The peripheral cells are also AB +ve, the central cells PAS +ve after Ox-AB-PAS-OG. In studying the precipitation of pituitary secretory granules by trichloroacetic acid (TCA), Leray (1966) found that while the acidophil granules were precipitated by lower concentrations of TCA, the central gonadotrops were not precipitated until the TCA concentration was raised to 10%, and the granules of the peripheral gonadotrops were only precipitated at higher concentrations, up to 15%; thus, the two gonadotrops in *Mugil cephalus* may be distinguished by their situation, staining reactions, and by the precipitation of their granules by TCA. The AT-PAS method also separates two gonadotrops in *Mugil capita*. Electron microscope studies indicate two types in the viviparous *Zoarces*. One type has granules 60-160 mμ diam, which tend to accumulate at one pole of the cell, and the endoplasmic reticulum consists of vacuolar cisternae; this cell becomes highly activated after parturition, in relation to ovarian follicular development. The second gonadotrop has granules 80-240 mμ diam, an endoplasmic reticulum with cisternae as dilated sacs or tubules, and is numerous and active during pregnancy.

At the ultrastructural level, Knowles and Vollrath (1966b) have given further details about the gonadotrops of the eel. The type with larger (190 mμ) granules may be an LH cell, and it persists with little change in seawater eels in which gonadal development was assumed to have started. However, the other type of gonadotrops, possibly an FSH cell, with granules of 130 mμ diam, apparently changes considerably as the eel becomes sexually mature; and Knowles and Vollrath (1966d) reported that they could identify typical FSH cells and, in addition, cells probably derived from these, with larger granules (up to 180 mμ), and with big empty vesicles and an active type of endoplasmic reticulum. In the trout, Follenius (1963a) described a cell type, probably a gonadotrop, which contained numerous granules of various sizes and showed extreme development of the endoplasmic reticulum in the form of smooth-walled vesicles. At certain stages of the activity cycle very large vacuoles appear in these cells.

From what has been written so far, the main tinctorial properties of the gonadotrops will be obvious. It is particularly interesting that PbH is so useful in distinguishing two types of gonadotrops, particularly when used in combination with PAS. This technique has also been valuable in study of the pars intermedia and ACTH cells, and it also has been used to separate two gonadotrops in mammals. More definitely histochemical investigations have been performed. Olivereau and Fontaine (1966) summarized their histochemical data on these cells in the mature eel: The PAS reaction indicates the presence of neutral glycoproteins, while the staining of the granules with Ox-AB at low pH (0.2) indicates their content of acid mucopolysaccharides, which is rather feeble however. The acidic nature of the gonadotrop granules is also indicated by their reactions with Gabe's AF after permanganic oxidation, and by their slight affinity for orange G in PAS-OG. In addition to these specific glycoproteinaceous secretory granules, the eel gonadotrops when mature also contain very large granules with variable tinctorial affinities (OG +ve, erythrosin +ve, PAS +ve, BA +ve, AF +ve, Aniline blue +ve), which may be lysosomes. Similar putative lysosomes occur in mature gonadotrops in *Poecilia*.

Olivereau (1962c, 1963b) found that eel gonadotrops are rich in SS/SH groups, a property they share with gonadotrops in other teleosts. In both the goldfish and *Poecilia sphenops*, the gonadotrops, and probably also the TSH cells (PAS +ve and AB +ve cells), incorporated ^{35}S more rapidly than other cell types, probably forming sulfate esters of acid mucopolysaccharides; but in the goldfish the gonadotrops only feebly incorporated labeled amino acids. According to Leray (1966) the gonadotrops in *Mugil cephalus* display only minor content of SS/SH groups, in contrast to other fishes. The same author reported that carp gonadotrops are rich in sialic acid and that the pituitary content of sialic acid increases with sexual maturation and falls after spawning. Sialic acid is a constituent of mammalian gonadotropins and TSH, and can be demonstrated histochemically in both the TSH and FSH cells of the cat.

Gonadotrops have been identified and described in many teleosts, usually in relation to the sexual cycle. It is usually accepted that any pars distalis basophils that are quiescent or absent before sexual maturity, and which show pronounced secretory changes in correlation with the gonadal cycle, are gonadotrops. In addition to these correlative studies, there are reports of characteristic changes in teleost gonadotrops following surgical gonadectomy or administration of gonadal steroids.

Gonadotrops of Poecilia latipinna

Like many cyprinodonts, the molly is viviparous, with approximately monthly cycles of gestation, parturition, and oocyte growth, and the young fish quickly reach sexual maturity, in a matter of a few months in males reared in good conditions. These facts make it an interesting species for studies on the gonadotrops. These cells were initially identified by comparing the pituitary in immature and mature fish. In the immature state, the proximal pars distalis contains only two chromophilic cell types, α cells and TSH cells, the latter identified by their responses to thiourea. On the ventral surface of this region, one finds many small chromophobic cells, with inactive nuclei, presumably undifferentiated cells. In the sexually mature fish, differentiated basophils are always found in this region, even during periods of sexual quiescence, when these cells assume an inactive appearance but are nevertheless recognizable as differentiated basophils. Thus, these ventral basophils would appear to be gonadotrops.

Confirmation of this identification comes from observations of changes in the ventral basophils during the monthly female cycle. The fully developed oocytes arc fertilized in the ovarian follicles by spermatozoa which may be stored embedded in the ovarian epithelium for long periods. The ovarian follicle acts as a brood chamber for the developing embryo. The fertilized egg is not ovulated but is retained in the follicle, and the embryo then develops on top of the yolky egg. Many embryos usually develop synchronously, forming a brood. When the fry are fully developed the follicles rupture, and the brood is expelled via the oviduct over a short period, a few hours at most. Under laboratory conditions, with 9 hr illumination per day and at 25°C, female fish undergo regular cycles of brood production, giving birth about every month, although the cycle is readily interrupted by unfavourable external conditions. During the greater part of pregnancy the ovary contains only previtellogenic (first phase) oocytes, along with the developing embryos; but at birth, or a few days earlier, some of these juvenile oocytes embark on rapid vitellogenic (second phase) growth and mature as a new crop of eggs ready for fertilization within 5-8 days of birth of the previous brood. Hypophysectomy experiments have shown that pituitary hormones are not essential for gestation or parturition, but are essential, as in oviparous forms, for the second phase oocyte growth. Since this phase is usually completed during the cycle within a few days starting at or just before parturition, one would predict that elevated pituitary gonadotropic activity is confined to this limited period of the cycle and is curtailed or absent during the much longer period occupied by pregnancy. Study of the gonadotrops tends to confirm this prediction, these ventral basophils displaying marked cytological and morphological changes correlated with the ovarian cycle. These changes are based on hitherto unpublished observations on scores of female *P. latipinna* together with systematic study and measurement on 30 fish sacrificed at known stages in the monthly cycle. The range of morphological alterations displayed by the gonadotrops in the cycle is impressive. As in many teleosts maximal activity in *Poecilia* gonadotrops is characterized by partial or total loss of the coarse glycoprotein granulation and development of vacuoles, in the presence of cytological evidence of marked synthetic and excretory activity (large nucleus and nucleolus, prominent Golgi image, and abundant cytoplasmic RNA). In addition to glycoprotein secretory granules, the active cells contain small numbers of larger amphiphilic granules or droplets, which have a varied staining affinity (positive to PAS, orange G, and erythrosin). They are negative to Alcian blue, however, so that in Ox-AB-PAS preparations these large magenta granules (R granules) contrast with the smaller dark blue secretory granules (B granules). The R granules are absent or very rare in quiescent cells, but they become numerous in the active state and are often to be seen lining the vacuoles that are characteristic of late vitellogenesis.

Orange G positive granules, no doubt equivalent to the R granules of *Poecilia*, have been described in gonadotrops of *Xiphophorus*, being particularly numerous together with vacuoles in the gonadotrops of gravid fish, presumably corresponding to the early pregnancy condition of *Poecilia*. The putative gonadotrops of *Oncorhynchus* contain large and small granules, the former being PAS +ve but AB –ve, while the small granules are positive to both PAS and AB; also in *Oncorhynchus*, the type 1 gonadotrops of Olivereau lose their glycoprotein granulation after spawning and develop vacuoles containing large OG +ve granules, a feature very like the appearance of R granules in *Poecilia* gonadotrops in late vitellogenesis. A similar phenomenon is seen in the gonadotrops of *Salmo salar*. The gonadotrops of *Tor (Barbus) tor* contain, in addition to glycoprotein granules, carminophilic granules that are particularly abundant in the vacuolated gonadotrops at the peak of their secretory activity. The development of vacuoles together with these acidophilic or amphiphilic granules and partial or total loss of glycoprotein granulation seems typical of the teleostean gonadotrop when highly active. As Olivereau (1967a) has suggested, the amphiphilic granules may be lysosomes; similar bodies occur in gonadotrops of amphibians, and lysosomes, with staining properties similar to those of these bodies in fish and amphibian gonadotrops, are known to be present in the basophils of the human pituitary. Presumably lysosomes would serve to rid the cell of excess secretory products, which may explain the marked development of R granules during early pregnancy in *Poecilia*, when oocyte growth (= gonadotropin secretion) is arrested, and also the similar granules in the postspawning gonadotrop of the salmon.

The question of whether *Poecilia* possesses two distinct types of gonadotrops is unresolved. The behaviour and staining properties so far described apply to all the cells in the gonadotropic zone, and PbH-PAS has so far been uninformative applied to *Poecilia*, all the gonadotrops staining with PAS and none selectively with the PbH. Sometimes it is possible to distinguish two apparently different cells with Ox-AB-PAS-OG at times of maximal gonadotropic activity, one type containing more R granules than the other, but in the absence of functional evidence for a clear separation there are no reasons for thinking that these are other than two stages in the activity cycle of one cell type.

The guppy, *Poecilia reticulata*, exhibits an ovarian cycle essentially like that of *P. latipinna*, and the gonadotrops also display a similar cycle, maximal activity of the gonadotrops being confined to a period starting in late pregnancy and finishing a few days after birth, the period when a new crop of oocytes is undergoing rapid vitellogenic development.

Little information is available about reactions of the gonadotrops to collateral changes in endocrine state. The eel gonadotrops were activated by treatment with an impure TSH, but this is probably to be attributed to the known contaminant, LH, since the gonadotrops of the eel are greatly stimulated by injections of human chorionic gonadotropin and carp pituitary homogenates or purified carp gonadotropin. It is unexpected that these exogenous gonadotropins, at the same time as they stimulate the eel gonad, should also cause activation of the eel pituitary gonadotrops; indeed, one might have expected that the gonadotrops would regress because of pituitary-gonad negative feedback. Probably a hypothalamic mechanism is involved, but the nature of this is not known. The quiescent gonadotrops of the immature eel are not affected by treatment with thyroxine or by radiothyroidectomy, injections of SU 4885, surgical interrenalectomy, sojourn in deionized water or seawater. Older work suggested that stress produced alterations in the gonadotrops of *Astyanax* and the goldfish resembling the changes following injections of ACTH or corticosteroids; such treatments generally lead also to inhibition of oocyte growth, suggesting failure of gonadotropin secretion. Atz (1953) presented in detail data showing that both ACTH and corticosterone administration in *Astyanax* led to transient inhibition of the gonadotrops followed by restoration, both at the beginning and end of the course of treatment; similar but less striking changes followed injections of saline twice daily, but not similar injections given once daily. These observations are probably best interpreted as a stress response, probably mediated by the ACTH-interrenal axis.

The gonadotrops regressed in ectopic pituitary transplants in *Poecilia formosa*, in correlation with evidence of failure of gonadotropin secretion.

Thyrotrops (δ cells or TSH cells)

The thyrotrops form the second division of the pars distalis basophils (mucoid cells) or the third division in those forms in which two gonadotrops have been demonstrated. In mammals, which display a corresponding trio of pars distalis basophils, a good deal of effort has gone into the tinctorial differentiation of the three types. Separation of the two gonadotrops is discussed by Herlant (1960, 1964), Purves (1966), and Carlon (1966, 1967). Separation of the TSH cells from the two gonadotrops is generally possible with a variety of staining techniques. In teleosts, it is generally found that the TSH cells closely resemble the gonadotrops in their staining properties, and most attempts to separate teleostean gonadotrops and thyrotrops purely by staining reactions have been inconclusive. However, there are exceptions. Barrington and Matty (1955) showed that in the minnow, *Phoxinus*, the gonadotrops are AF –ve but the thyrotrops are AF +ve; Matty and Matty (1959) confirmed this distinction in *Rutilus*, and described in other species both AF +ve and AF –ve basophils, without functional identifications. Similarly, in *Caecobarbus geertsi*, the putative thyrotrops are AF +ve, but the putative gonadotrops are AF –ve. More subtle colour distinctions occur in other teleosts, for example, in *Astyanax mexicanus* where both thyrotrops and gonadotrops stain with AF, but the gonadotrops stain more intensely by PAS. In the eel, the thyrotrops are more strongly AF +ve and AB +ve than the immature gonadotrops, and take on a slightly different colour after PAS-OG. Tinctorial distinction between thyrotrops and gonadotrops is also possible in *Poecilia*, in which the TSH cells stain a paler blue with Ox-AB at pH 0.2, and the gonadotrops, when active, contain many R granules which are present in small numbers, if at all, in the TSH cells. With Aliz B, the gonadotrops stain a deep blue, but the thyrotrops take on ii slatey gray-blue.

Fortunately, the visual distinction between TSH cells and gonadotrops depends not only on such tenuous tinctorial reactions, but also on marked topographical separation of the two cell-types. Reference has already been made to the fact that the TSH cells lie in the rostral pars distalis, as in the eel, or in the proximal pars distalis, as in *Poecilia*. Even in the latter type, when both TSH cells and gonadotrops occur in the proximal pars distalis, there is usually a spatial separation between them, the TSH cells lying dorsally (e.g., *Poecilia* and other cyprinodonts) or antero-dorsally (cichlids and *Mugil*), while the gonadotrops tend to lie in the ventral region, sometimes forming a virtual gonadotropic zone as in *Poecilia*. However, this special separation may not always be clear (e.g., *Astyanax*).

In the immature male eel, the TSH cells are small angular elements, lying in the rostral pars distalis between the follicles of η cells, and sometimes infiltrating into the follicle walls, with very fine secretory granules staining with Ox-AB, Aniline blue, thionine, PAS, and AF. In immature female eels, the TSH cells are similar, but here they tend to form large cell cords between the follicles in the middle of the rostral region.

In *Poecilia* the thyrotrops lie in the finger-like inner prolongations of the proximal pars distalis into the neurohypophysis, mixed with the more numerous a cells. They are generally rounded or angular cells, more or less square in section, with a large rounded nucleus and a rather coarse granulation, positive to PAS, OxAB, AF, and Aniline blue.

In all teleosts in which the thyrotrops have been identified, they display the typical staining reactions of basophils. Surprisingly little information seems to be available specifically about the histochemistry of the cells. In immature *Mugil*, the thyrotrops incorporate labeled cysteine, and they also incorporate $^{35}SO_4Na_2$ in *Poecilia sphenops*, although more weakly than the gonadotrops. Correspondingly the TSH cells of the guppy contain SS and SH groups, although Olivereau (1962a) could find no evidence of these groups in the thyrotrops of the eel. Certain cells in the rostral pars distalis of the goldfish, identified by Olivereau (1962b) as TSH cells, rapidly incorporated glucose-^{3}H, probably an indication of their synthesis of mucopolysaccharides and glycoproteins. Otherwise, one must infer the histochemistry of the TSH cells from consideration of their general staining reactions, as done by Olivereau and Fontaine (1966) for the eel gonadotrops.

The TSH cells of the juvenile river eel are revealed by the electron microscope as containing small vesicles or secretory granules (ca, 140 mμ diam), and a diffuse endoplasmic reticulum with wide cisternae; in silver eels caught on their migration to the sea, the endoplasmic reticulum was more prominent and the secretory granules more concentrated, suggesting greater activity. The presumptive TSH cells of *Zoarces viviparus*, in contrast, contained large granules (diam about 400 mμ), with foamlike cisternae in the endoplasmic reticulum; in some cases, smaller granules of varied size (120-160 mμ) could be found in these cisternae. The trout TSH cells displayed a well-developed endoplasmic reticulum, with numerous smooth-walled cisternae. The secretory granules varied greatly in size, from 100 to 800 mμ.

Nearly every student of the teleost pituitary has described basophils in the pars distalis, usually in the proximal region, but in some species in the rostral region also; however, most workers have not distinguished TSH cells from gonadotrops either tinctorially or functionally.

Evidence for the secretion of TSH by the δ cells

The most detailed investigation of pituitary—thyroid relationships in teleosts is that of Olivereau on the eel. Massive doses of radioiodine resulted in partial or total destruction of the thyroid in male silver eels. The animals were sacrificed at different times after the radioiodine treatment, up to 7 months, and even at 1 month after radiothyroidectomy, the δ cells were degranulated and hypertrophied, and displayed mitotic activity. The cells enlarged, losing their angular outline and the cytoplasm became foamy and vacuolated, and the nucleus and nucleolus hypertrophied. Although the η and ε cells were slightly affected in this experiment, the reactions of the δ cells were so strong and characteristic that there can be little doubt of their thyrotropic function; this marked degranulation of these cells in the radiothyroidectomized male eel is paralleled by a fall of nearly 80% in the TSH content of the pituitary of radiothyroidectomized female eels. The complement of radiothyroidectomy was the treatment of immature male eels with thyroxine, which produced marked histological signs of involution in the thyroid gland, accompanied by inactivation changes in the δ cells, which shrank and lost their granulation, finally becoming small inactive chromophobic cells with small nuclei and nucleoli. As in the case of the radiothyroidectomy experiment, collateral effects on the η cells and ε cells were observed, but the pronounced changes in the δ cells marked them as the thyrotrops.

In the trout, *Salmo gairdneri*, iodine deficiency stimulated the thyroid gland and led to an increase in the number of δ cells and in the size of their nucleoli; subtotal radiothyroidectomy resulted in strongly hypertrophied δ cells, with enlarged nuclei and nucleoli and loss of glycoprotein granulation, again identifying these elements as thyrotrops. Subtotal radiothyroidectomy of the goldfish produced results in agreement with those in eel and trout, the δ cells becoming strongly activated. Further work on the goldfish demonstrated that partial or total radiothyroidectomy led to reduced thyroidal ^{131}I uptake and transformed the granulated δ cells into hypertrophied and hyperplastic chromophobes with large nuclei and nucleoli.

Using *in vitro* techniques, Baker has extended and confirmed the identity of the TSH cells in eel and trout. When isolated from the hypothalamus in culture, trout δ cells degranulate fairly rapidly and gradually assume an appearance of increased activity. The addition of thyroxine to the culture medium specifically prevents this rapid loss of δ granules and leads more slowly to regression of the cells, which become smaller and exhibit nuclear and nucleolar atrophy. The eel delta cell *in vitro* degranulates more slowly and maintains an appearance of activity; addition of thyroxine to the medium from the start of culture results in nuclear atrophy. If trout glands are transferred from a thyroxine medium (granulated δ cells) to a control medium, the δ cells rapidly degranulate, showing that inhibition of their discharge by thyroxine is reversible; however, if the reverse transfer is made after the δ cells have been allowed to degranulate on the control medium, thyroxine is unable to regranulate the cells. In both eel and trout, δ cells maintained on thyroxine medium, with histological signs of regression, display impaired incorporation of uridine-^{3}H into the nucleus compared with cells on control medium. Exposure to thyroxine had no effects on any of the other pituitary cell types, confirming that the δ cells in both species are thyrotrops, and that thyroxine-TSH feedback appears to operate mainly at the level of the pituitary cell.

The δ cells of *Poecilia* are generally rather few in number and lie associated with the α cells. Their thyrotropic nature is indicated by their early differentiation, since they are the only pars distalis basophils in newly born fish, the gonadotrops only becoming differentiated much later. Taking immature fish (6 months old or less), with undifferentiated gonadotrops, Ball (1987) has studied the response of the δ cells to treatment with the antithyroid agent thiourea, administered in solution in the tank water. Results are clearly demonstrate the strong response of the δ cell to thiourea-induced blockage of thyroid hormone synthesis. None of the other cell types responded to the treatment. Unexpectedly, treatment of adult fish with thyroxine produced no obvious regression of the δ cells, although reducing thyroidal 24-hr ^{131}I uptake to 0.5%. This lack of effect of thyroxine on TSH cell morphology in *Poecilia latipinna* agrees with the failure of thyroxine treatment to affect the TSH cells of the guppy, but contrasts with the distinct effects in the eel and *Mugil*, and with the rather slight changes in the thyroid and thyrotrops of *Astyanax*. Clearly the time course of the TSH cell response to different doses of thyroxine should be explored in *Poecilia*.

Leray and Blanc (1967a,b) showed that treatment of young *Mugil auratus* with the thyroid inhibitor propylthiouracil (PTU) reduced levels of plasma protein-bound iodine (PBI)—that is, impaired the production of thyroid hormone—and produced signs of histological stimulation in the thyroid itself (i.e., raised levels of TSH). Correspondingly, the δ cells of the pars distalis became strongly activated, marking them as thyrotrops since no other pituitary cells were affected. Olivereau (1968) confirmed these observations in this same species, thiourea activating and thyroxine inhibiting the δ cells.

Other evidence bearing on the function of the δ cells comes from their marked activation in response to thiourea or thiouracil treatment in *Phoxinus*, *Astyanax*, and the guppy. Thyroxine administered with thiourea to the guppy reversed the activation of the δ cells induced by thiourea alone. Confirmation comes from Schreibman's work (1964) on *Xiphophorus*, in which the δ cells, centrally located in the proximal pars distalis and differentiated very early in life, were strongly hypertrophied and hyperplastic in fish with thyroid tumors. The cavefish, *Caecobarbus*, has a quiescent thyroid, and also displays very inactive-looking δ cells.

Reference has already been made to the early differentiation of the TSH cells in *P. latipinna* and *Xiphophorus*, long before the gonadotrops are differentiated, and these cells also differentiate early in the guppy and the cichlid *Herichthys*. The TSH cells sometimes exhibit functional changes during the life history. They are present only in small numbers in trout and the freshwater (parr) stages of the salmon, *Salmo solar*, in correlation with the very low content of TSH in these glands; however, in the salmon smolt, with a highly active thyroid, and about to migrate to the sea, TSH cells are numerous and highly active. In *Xiphophorus maculatus*, Schreibman (1964) described senility changes in aging females, in which the TSH cells became smaller, with smaller nuclei, and exhibit degranulation and vacuolization; in the related guppy, thyroid activity decreases with aging, which supports interpretation of the senescence changes in the TSH cells of *Xiphophorus* as indicating reduced secretion of TSH. In young *Xiphophorus* the TSH cells look much more active. Silver eels exhibit vacuolization and ultrastructural features of their TSH cells which perhaps indicate a greater activity than in the yellow stage.

Some collateral endocrine changes have been found to affect the TSH cells. They were slightly activated after castration in *Xiphophorus*, and another slight suggestion of gonadal influences on the TSH cells perhaps comes from Olivereau's finding (1967a) that the cells are slightly less active in immature female eels than in immature males (although we do not know that the eel gonads at this stage secrete sex hormones). In the case of eels brought to sexual maturity by artificial treatment, the reactions of the TSH cells are difficult to interpret. When maturation was induced by injections of TSH contaminated with LH, the regression of the eel TSH cells is probably a response to thyroidal activation induced by the exogenous TSH; however, in male eels brought to maturity by human chorionic gonadotrops, the TSH cells were activated in parallel to activation of the thyroid, possibly the result of some unknown interplay between the exogenous gonadotropin and the hypothalamus. The inactivation of these cells in female eels matured by injecting carp pituitary extracts is in part at least attributable to carp TSH in the extracts. It is perhaps worth recalling that the question of whether or not sexual maturation in teleosts is *necessarily* accompanied by thyroidal hyperactivity is still open, with conflicting reports in the literature.

Eel TSH cells were not modified 8 or 9 days after surgical interrenalectomy but were quite strongly activated after 2-5 days treatment with the adrenocortical inhibitor SU 4885, and SU 4885 also activates the TSH cells after 5 days' treatment in *Poecilia*, although not at 2 days.

An unexpected reaction of the TSH cells of the eel occurs in response to injections of ovine prolactin. Olivereau (1966b) found that chronic injections of prolactin resulted in marked histological activation of the thyroid in intact eels but not of hypophysectomized animals. After only 2 or 3 daily injections, the TSH cells in the intact animals were slightly hypertrophied, with occasional mitoses, and with further injections the TSH cells in the intact animals became extremely activated and displayed many mitoses. As Olivereau pointed out, the effects of the exogenous prolactin could result from its acting as a goitrogen, preventing the formation of thyroid hormone and, hence, by removing feedback inhibition, resulting in elevated TSH output; or, the ovine prolactin could somehow be causing elevated TSH output even in the presence of a normally functional thyroid, as if it were acting like a TSH-releasing factor. A similar action of ovine prolactin on the TSH-thyroid axis in the amphibian *Rana catesbiana* appears to represent a goitrogenic effect, although this is not apparently true for thyroidal stimulation by prolactin in a urodele. Subsequent work on the eel showed that prolactin does not act like a goitrogen in this case, since in addition to the histological signs of thyroidal activation, prolactin also increased ^{131}I uptake by the gland; thus, the activation of the eel TSH cells by ovine prolactin presumably depends on some hypothalamic or intrapituitary mechanism not yet understood.

Atz (1953) observed striking changes in the thyrotrops of *Astyanax* following injections of cortisone or ACTH daily for 10 days. The TSH cells underwent a cycle of degranulation and vacuolation followed by regranulation, both at the beginning and the end of the injection period, and also displayed mitotic activity. The fact that similar effects were produced by daily saline injections suggests that Atz was observing a response to stress normally mediated by the ACTH-adrenal axis.

The TSH cells are better granulated in eels in seawater than in freshwater animals, probably indicating a lower rate of secretion in seawater related to the greater availability of iodine in this medium. However, Knowles and Vollrath (1966c) reported that TSH cells in some eels caught at sea were more numerous than in river eels although in some seawater eels they could find no typical TSH cells. Silver eels during the downstream migration displayed very active TSH cells, judging by their extremely prominent endoplasmic reticulum. Thus, even in the one species, the relationship between TSH cell activity and external salinity is unclear. The eel TSH cells possibly tend to be less active in deionized water than in freshwater, although this was not a very marked reaction.

Reference was made earlier to the fact that in both eel and trout pituitaries, the TSH cells retain their activity *in vitro*, the trout cells actually degranulating and becoming apparently more active than *in vivo*. Comparable data comes from ectopic pituitary homotransplants in *Poecilia formosa* which secreted TSH at normal levels, or in some individuals perhaps higher than normal levels, associated with the retention of very active TSH cells in the transplants, often degranulated but with very active nuclei. Comparable retention of TSH function and of active TSH cells has been observed in ectopic pituitary autotransplants in *P. latipinna*. The implications of these findings for the analysis of hypothalamic control of TSH function in teleosts are obvious.

Pars intermedia

The pars intermedia partially or completely surrounds the distal and largest region of the neurohypophysial core and is usually extensively invaded by neurohypophysial tracts (exception: *Lepidogobius*). Its relative size varies in different species. It may be the smallest region of the adenohypophysis, as in *Poecilia* and other cyprinodonts, or, together with the enclosed neurohypophysis, it may form as much as two-thirds the volume of the whole pituitary, as in salmonids. Cells from other regions of the adenohypophysis may invade the pars intermedia in small numbers (e.g., *Hippocampus*), and pars intermedia cells are sometimes found in the pars distalis.

Two cell types may usually be distinguished by their staining reactions and often also by differences in shape and position. Staining characteristics vary with species; pars intermedia cells have been described as basophilic, staining for example with Aniline blue, or acidophilic (with azocarmine or orange G), or amphiphilic, taking both the blue and orange components in trichrome and tetrachrome techniques. Some species have been described as having only one chromophilic cell type in this region, together with chromophobes (e.g., the guppy, *Oncorhynchus*), or all the cells in the pars intermedia may appear chromophobic (*Rutilus*, *Carassius*). In such cases, it is impossible to be certain that two cell types exist, although probably, as in the case of *Carassius*, more detailed investigations would reveal this to be so. Within any one species the staining properties may vary, depending on the physiological conditions of the cells and also on the techniques applied. Thus, in *Perca fluviatilis*, the PAS +ve cells bordering the neurohypophysis have been described as basophilic or acidophilic. A more precise differentiation between two pars intermedia cell types is achieved by staining with PAS followed by lead hematoxylin (PbH). The resulting differentiation was first described in Mugilids by Stahl (1958): One cell type, oval in shape and bordering the neurohypophysis, was PAS +ve, and a second type, lying further from the neurohypophysis but connected to it by cellular prolongations, was PAS –ve but PbH +ve. Several teleostean pituitaries have since been examined by this technique, and generally the two types of cells have been differentiated. The PbH +ve cells frequently have a "club" shape as in *Mugil*, and they are often chromophobic by classic techniques (e.g., *Carassius*, *Poecilia*).

Two cell types are not invariably present in the pars intermedia; for example, the PAS +ve cell appears to be missing from salmonids, in which all the cells look alike and stain with PbH. However, most species studied display a PAS +ve cell type, often together with a chromophobe that no doubt represents the PbH +ve cell. Although PAS and PbH appear to be mutually exclusive in staining the two cell types, there seems to be no correlation between the reactions of the cells to trichrome stains on the one hand and to PAS on the other; thus, the PAS +ve cells may be basophilic as in *Hippocampus*, *Phoxinus*, and *Anguilla*, or acidophilic as in *Cichlasoma*, *Blennius*, and *Zoarces*, or amphiphilic as in *Poecilia latipinna*. It is our impression that the reaction of these cells in mixed staining techniques is more than usually sensitive to slight variations in timing, strengths of dye solutions, etc. Similarly the PbH +ve cells have been variously described as chromophobic (most often), acidophilic, or weakly basophilic.

Instrusive islets of the amphiphilic PAS +ve cells occur in the pars distalis of *Poecilia latipinna* and *P. formosa* particularly in the rostral region. The number of such islets increases with age. Similar islets of PAS +ve cells, presumably pars intermedia cells, are found in the rostral pars distalis of other cyprinodonts. The PbH +ve cells have not been found outside the pars intermedia in *Poecilia*.

Pars intermedia histochemistry has been neglected. The PAS +ve cells of *Poecilia sphenops* incorporate $^{35}SO_4Na_2$ rapidly, and in *Mugil* their granules are precipitated by a much lower concentration of trichloroacetic acid (0.5%) than the granules of any other cell type in the gland.

Acidophilic droplets, apparently colloidal, often occur between the pars intermedia cells. Similar material in the pars intermedia of other vertebrates has often been interpreted as an accumulation of hormonal material or as the products of cellular degeneration.

Few ultrastructural studies have been made on this region. The presence of two distinct cell types is confirmed in most studies, and in *Anguilla* a third cell type has been reported, but it is not clear what this corresponds to in light microscope studies. With the electron microscope, the two main cell types are most readily distinguished by differences in size of their osmiophilic secretory granules, although mitochondrial size and details of the endoplasmic reticulum are other criteria. Granule size and shape differ in different species and do not correlate with granule staining reactions. Thus, in *Zoarces* the PAS +ve cells have smaller granules (160-200 mμ) than the PbH cells (up to 400 mμ), and the same is true of the eel (1.80 mμ vs. 320 mμ). In contrast, in *Hippocampus* the PAS +ve cells have the larger granules

(250-300 mμ vs. 200 mμ). However, as in pars distalis cells, the granule size may vary under different conditions even within one cell type; for example, in the elver of *Anguilla* the average granule size is smaller in both cell types than in the older freshwater eel. The significance of these differences is unknown.

Follenius (1963a, 1965a), on the basis of his ultrastructural observations on the guppy, *Perca* and *Salmo*, considers that the two cell types of the pars intermedia are not distinct, but probably represent different stages in the functional cycle of a single cell type, an interpretation at odds with the conclusions of most workers at the light microscope level; the independence of the two cell types is indicated by their different reactions in the eel kept in deionized water.

Evidence for the functions of the pars intermedia cells

To understand the difficulties attending any allocation of function to the pars intermedia cells, it is necessary to appreciate that there is at present no general agreement about the total physiological role of this part of the gland. By homology with the tetrapod pars intermedia, it is to be expected that this region is the site of synthesis of melanophore-stimulating hormone (MSH, intermedin), and there is now good experimental evidence to support this idea. Extracts of teleostean pituitaries affect the distribution of pigment in both the melanophores and erythrophores of fishes, and experiments with separate extracts of anterior and posterior halves of the gland show MSH activity to be most abundant in the pars intermedia region. Long-term *in vitro* culture of anterior and posterior halves of the trout pituitary showed that 11-12 times more MSH is produced by the pars intermedia than by the rest of the gland; the small amount of MSH secreted *in vitro* by the anterior region may be attributed to the inclusion of some pars intermedia cells in these fractions.

The existence of two distinct cell types suggests the production of two distinct hormones by this region. On other grounds, it has been proposed that two antagonistic melanophorotropic hormones are secreted by the fish pituitary, one (MSH) causing melanophore dispersion, the other (melanophore-concentrating hormone, MCH) inducing melanin concentration in the middle of the pigment cell. Stahl (1958) suggested that the two pars intermedia cell types might produce these two factors. This two-hormone hypothesis is not universally accepted, and some workers consider the experimental data, which originally led to the two-hormone concept in amphibians and fishes, are explicable in terms of only one hormone.

This lack of certainty about whether the pars intermedia produces one or two melanotropic factors is reflected by uncertainty about the functions of the two cell types of this region: At present, if only one hormone is secreted by the pars intermedia we cannot say which cell produces it; and if the other cell type does indeed secrete a separate hormone, we cannot be sure of its functions, except to say that there are some indications that such a second pars intermedia factor might have function(s) quite different from pigmentation control. The conditions under which the pars intermedia cells show cytological evidence of functional alterations are, therefore, of quite unusual interest.

Attempts have been made to locate the cellular origin of MSH by observing cytological changes in the pars intermedia in response to adaptation to a black or white background. In *Cichlasoma* and *Blennius* adapted to a black background for several weeks, the PAS +ve cells became larger and better granulated, with larger nuclei and nucleoli than in fish maintained on a white background, in which these cells regressed. While these observations point to the secretion of MSH by the PAS +ve cells, other work produced conflicting data. Thus, Knowles and Vollrath (1966a) kept eels for several weeks in complete darkness, and then transferred them to an illuminated black or white background for 4 hr before sacrifice. There were many PAS +ve cells in the pars intermedia of animals transferred to the black background, and in these eels the melanophores were in a dispersed state, indicating high levels of MSH secretion. In contrast, the eels transferred to a white background displayed concentration of melanophores (i.e., low levels of MSH secreted), and in the pars intermedia few PAS +ve cells could be found. Since in such a short-term experiment the scarcity of PAS +ve cells in the white-background fish must have been owing to discharge of secretory granules, we would interpret these observations as indicating that MSH is not secreted by these cells. Unfortunately, in this work there is no detailed information about the state of the pars intermedia in the eels in total darkness, but it seems that they displayed PAS +ve cells in the pars intermedia, which lends support to our interpretation. In other teleosts (*Ameiurus*, *Carassius*, and *Spinachia*), adaptation to black or white backgrounds did not lead to cytological alterations in the pars intermedia. Chavin (1959), however, reported an increase in MSH content and hypertrophy of the amphiphils in the pars intermedia of black-adapted *Carassius*, which would indicate that these cells, PAS –ve and PbH +ve, secrete MSH. However, this interpretation is not entirely secure since Chavin stated that the gonadal state of his fish was variable, and there does exist some evidence of a correlation between the number and size of these PbH +ve cells in *Carassius* and the degree of gonadal maturation. Nevertheless, the different MSH contents of the pituitary of black-adapted and white-adapted fish is interesting and highly suggestive, and similar experiments should be repeated to avoid the complication of gonadal variations.

Melanin dispersion in fishes may occur independently of background adaptation. It should perhaps be emphasized that teleostean melanophores are under dual control, neural and humoral, and in some species (e.g., *Anguilla*) humoral control predominates, so that even innervated melanophores will respond to MSH, while in other species (e.g., *Fundulus* and *Poecilia*) the neural component predominates and the melanophores must be denervated to demonstrate experimentally their response to MSH. In both *Anguilla* and *Poecilia*, the adrenocortical inhibitor SU 4885 causes melanin dispersion, and in *Anguilla* (but not in *Poecilia*) this is accompanied by degranulation and hypertrophy of the PbH +ve cells, the PAS +ve cells remaining unchanged. However, in both species the ACTH cells were activated by the drug, and since mammalian ACTH possesses intrinsic MSH-like properties, the melanophore response in *Anguilla* could be an effect of endogenous ACTH. Furthermore, the darkening in *Poecilia* in response to SU 4885 involved innervated melanophores, which in this species are not directly responsive to exogenous MSH or ACTH, and it is probable that some narcotic property of the drug was responsible for this effect in *Poecilia*, if not, indeed, in *Anguilla*.

In Salmo salar, the cells of the pars intermedia were inactive in the parr during the winter months, became more active in the spring, and were extremely active during the smolt stage (i.e., the stage of migration to the sea), undergoing intense mitotic activity so that the whole region enlarged. Olivereau (1954) correlated this with the rapid deposition of melanin pigment in the smolt pectoral fins. In the adult salmon migrating upstream from sea to spawning grounds, the pars intermedia appeared to be degenerate, but was restored in spawning and postspawning fish, although not as active as in the smolt. The blind cavefish, *Caecobarbus geertsi*, is characterized by a total lack of melanin pigmentation, and Olivereau and Herlant (1954) found in correlation that the cells of the pars intermedia were inactive and "embryonic" in character.

Increased activity of the pars intermedia and increased melanin dispersion occur in *Anguilla* during a prolonged period in deionized water, the whole pars intermedia becoming hypertrophied. The PAS +ve cells undergo a rapid and intense degranulation and hypertrophy, while the PbH +ve cells only degranulate slowly. Since melanin dispersion also develops slowly, Olivereau tentatively suggested that the PbH cells might be the source of MSH. This conclusion would disagree with the results already cited from background adaptation of *Cichlasoma* and *Blennius*. Some role of the pars intermedia in electrolyte regulation may be inferred from these results of Olivereau, and it is worth recalling that osmotic stress is often associated with changes in the pars intermedia of tetrapods. More recently, Olivereau (1968) has reported that gradual transfer of *Mugil auratus* from seawater to freshwater results in activation of the PAS +ve cell in the pars intermedia, the PbH +ve cell remaining in its normal state of high activity; there are no data about pigmentation in these conditions, but the results perhaps support those on the eel in suggesting some nonmelanophorotropic function for the PAS +ve cells, again in opposition to the results from the background adaptation experiments of Baker and Chavin. Another confusing observation is that the PAS +ve cells in the pars intermedia of the elver (young eel) are activated during the upstream migration from the sea; since at the same time as they move into freshwater these animals also become more heavily pigmented, it is impossible to know whether this indicates an osmoregulatory or a pigmentary function for the PAS +ve cells.

A variety of observations seems to relate the pars intermedia to reproductive changes. For example, in *Zoarces viviparus* the PAS +ve cells increase in number and activity during oogenesis and during the breeding season, and in *Carassius* the PbH +ve cells become more numerous during gonadal maturation while the PAS +ve cells increase after the reproductive period. In the herring, *Clupea*, intense secretory activity has been seen in the pars intermedia during the spawning period. It is not yet possible to offer functional interpretation of such observations, but the variety of conditions affecting the activity of the pars intermedia cells underlines the possibility that the hormone(s) produced by this region might be concerned in functions other than control of pigmentation.

Both trout and eel pituitaries will synthesize and release MSH *in vitro* after separation from the hypothalamus. In the case of the eel, a cytological study of the cultured glands has failed to determine the source of MSH, since both the PAS +ve and the PbH +ve cells underwent gradual degranulation but retained an active appearance. In the trout gland *in vitro*, the rapid decline in the stored MSH content of the pituitary is associated with degranulation of the single PbH +ve cell type, in the presence of active-looking nuclei and nucleoli. In contrast, both cell types in the pars intermedia of ectopic pituitary transplants in *Poecilia formosa* presented a histological picture of involution.

Neurohypophysis in Teleosts

The neurohypophysis is composed of axonal nerve fibers, mostly nonmyelinated, originating from neuronal cell bodies in the hypothalamus. These nerve fibers extend down the pituitary stalk (neurohypophysial tract) into the pituitary gland to form the main bulk of the neurohypophysis (neurohypophysial core) which forms the center of the whole gland. Many fibers terminate in the neurohypophysial core in close relationship to blood vessels, and many other fibers extend beyond the core and invade all regions of the adenohypophysis, an arrangement peculiar to teleosts. Scattered ependymal cells and glial cells, collectively termed "pituicytes," occur throughout the neurohypophysis but are especially abundant in the posterior part of the core close to the pars intermedia. Tubelike extensions of the third ventricle bordered by pituicytes may penetrate into the posterior neurohypophysial core, forming a hollow center in the fingerlike projections of the neurohypophysis into the pars intermedia.

A special feature of the fish pituitary is the extensive penetration of the pars intermedia by the neurohypophysis, the two regions mixing in complex interdigitation. Teleosts have specialized further in having the anterior part of the neurohypophysis penetrate the pars distalis with fingerlike processes, although in a less intricate manner than in the pars intermedia. This ramification of the neurohypophysis into the adenohypophysis is not present in salmonids at hatching, and thereafter it develops slowly and is only very slight even at 4 months in the trout. Similarly in the elver, there is only slight indication of the extensive penetration of the pars intermedia by neurohypophysial processes, so characteristic of the adult eel.

Two types of neurohypophysial fibers may be distinguished with the light microscope, depending on whether or not the material contained in the fiber will stain with the classic neurosecretory stains, AF, chrome alum hematoxylin (CAH), aldehyde thionin (ATh), and AB. The two fiber types appear to correspond to the Type A ("stainable") and Type B ("nonstainable") fibers distinguished at the ultrastructural level by Knowles. Neurosecretory material in both "stainable" and "nonstainable" fibers may accumulate in the neurohypophysis in large masses termed "Herring bodies."

The neurohypophysis and the neurosecretory cells of the hypothalamus (mainly but perhaps not all, collected into groups termed "hypothalamic nuclei") thus form a functional unit concerned in the synthesis transport and release of neurosecretory materials. They also display, anatomically at least, the fundamental feature of neurosecretory elements, in that they form "a final common pathway linking the nervous and endocrine systems".

Various octapeptides with characteristic biological properties have been isolated from the pituitary in all vertebrate groups and in many cases have been shown to be associated with the hypothalamus and neurohypophysis. It is these octapeptides which are believed to be synthesized in the hypothalamic nuclei and passed down to the neurohypophysial core, probably contained in or associated with ultrastructural neurosecretory granules, which are generally considered equivalent to the stainable material seen with the light microscope. In teleosts, two octapeptides have been found in the pituitary, arginine vasotocin (AVT) and isotocin (IT), and these are thought to be the neurohypophysial principles of these fishes.

"Stainable" fibers, probably Knowles' type A

The stainable fibers probably mostly originate from the nucleus pre-opticus (NPO) of the hypothalamus, and they contain neurosecretory material that stains with AF, CAH, ATh, and AB. The majority of these fibers probably terminate in the distal part of the neurohypophysial core surrounded by the pars intermedia, ending in association with pituicytes or blood vessels or with the pars intermedia cells. In addition, fibers with these staining reactions have been observed closely associated with cells of the pars distalis. In the eel, Olivereau (1967a) distinguished two kinds of neurosecretory fibers with the light microscope; one type containing classic neurosecretory material and mainly passing toward and penetrating the pars intermedia, the other type with material PAS +ve but AF –ve, which penetrates the pars distalis. She tentatively equated these with Type A and Type B fibers originating from the NPO and NLT. Similar distribution of the two kinds of fibers can be seen in *Poecilia*.

After hypophysectomy, classic neurosecretory material accumulates along the course of the severed nerve axons in the infundibular area. Transection of the NPO-pituitary tracts in *Lepidogobius* also led to accumulation of AF +ve material proximal to the cut, and eventually to loss of this material from the neurohypophysis. Thus the neurosecretory material is certainly formed in the hypothalamus and transported to the pituitary along the axons, as in other vertebrates. Exposure of hypophysectomized *Porichthys* to continuous light induced movement of AF +ve material from the NPO to the cut ends of the axons in the infundibulum. The ends of the severed axons in hypophysectomized goldfish and *Porichthys* regenerate so that the severed neurohypophysial stalk eventually forms a kind of isolated neurohypophysial core with axons terminating on blood vessels as in the normal condition. In hypophysectomized *Clevelandia*, the neurosecretory material that accumulated proximal to the cut ends of the axons eventually disappeared, probably into blood vessels that associated with the fibers.

The correspondence of this light microscope category with the Type A fibers defined by ultrastructure is almost certain, though perhaps not definitively established. Knowles (1965) defined Type A fibers as containing osmiophilic granules more than 100 mμ in diameter, in contrast to Type B fibers with smaller granules; and in *Perca* and *Salmo* Type A fibers contain granules the same size as those in the cell bodies of the NPO. Several kinds of Type A fibers have been described differing in the size of their granules, their destination within the pituitary, and in their response to experimentation.

Within the pituitary of *Anguilla* and *Conger*, the Type A fibers mostly terminate in or close to the pars intermedia, or on extravascular channels between the endocrine cells, or make direct contact with the pituicytes, or occasionally with the pars intermedia cells. A Type A3 fiber has been distinguished in the eel pars distalis and appears to be distinct from the A1 and A2 fibers found in the neurohypophysial core close to the pars intermedia. The Type A3 fibers in the pars distalis of the eel, like the more numerous Type B fibers in this region, terminate on extravascular channels, very close to the intrinsic endocrine cells.

Light microscope accounts of changes in the neurosecretory material in the neurohypophysis nearly all relate to the Type A fibers in the distal neurohypophysial core, close to the pars intermedia. The amount of stainable material in this region commonly alters in response to osmotic stimuli. In many teleosts, transfer to a hypertonic medium results in loss of this material, both from the neurohypophysis and also from the NPO. Eels adapted to seawater have less stainable material in the neurohypophysial core than freshwater eels; on the other hand,

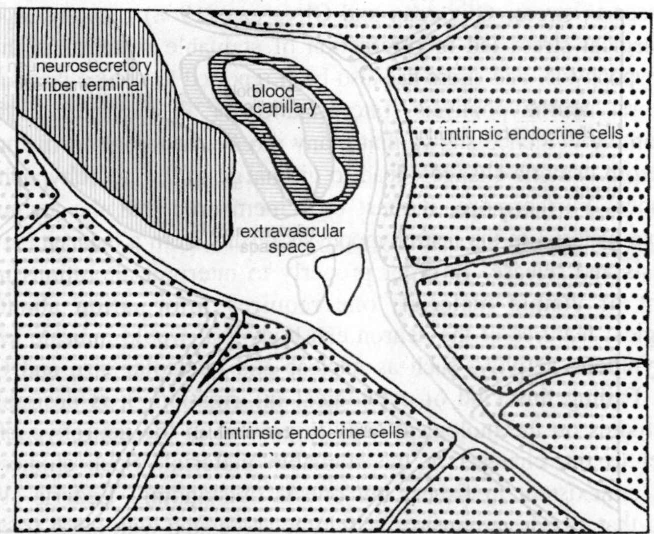

Fig. 2.2. Diagram to illustrate the relationship between a neurosecretory fiber terminal, intrinsic endocrine cells, and a blood capillary in the proximal pars distalis of the eel.

the total amount of stainable material in the hypothalamo-neurohypophysial tracts did not appear to alter in response to changing salinity in the eel or the goby, *Clevelandia ios*, nor did the material in the neurohypophysial core change appreciably in *Mugil* subjected to progressive reduction in salinity. Gasterosteus transferred from one-third seawater to freshwater exhibited a rapid loss of stainable material, while in hypotonic media, or in response to dehydration, stainable material increased in the pars magnocellularis of the NPO of *Misgurnus*, but decreased in the parvocellular elements, suggesting different functions for the two components of the NPO.

In the neurohypophysis of eels kept for long periods in deionized water, Olivereau (1967b) found that stainable material tended to increase at first but then decrease, the neurohypophysis being compressed by hypertrophy of the pars intermedia.

Changes such as these have usually been interpreted as indicating release or retention of neurohypophysial factors in response to osmotic demands, there being some experimental evidence that neurohypophysial factors, especially arginine vasotocin, may be concerned in ionic regulation and water balance in teleosts. However, in some of the experiments in which fish were transferred between different salinities, the osmotic conditions imposed must have been very severe, and it is possible that the neurohypophysial changes could have been nonspecific responses to stress. Olivereau and Olivereau (1968) recently reported that surgical stress led to rarefaction of stainable material in the eel neurohypophysis, and stress has also been reported to reduce the material in the rat neurohypophysis. Interpretation is all the more difficult, since ample evidence shows that many stress conditions in mammals, apart from osmotic changes, lead to discharge of the neurohypophysial hormones. Furthermore, it must be remembered that the amount of stainable material in the neurohypophysis results from a balance between synthesis and release. In order properly to interpret alterations in the amount of visible material, one requires information about the cytological features of the neuron cell body that would indicate low or high synthetic activity, such as nuclear and nucleolar size and Golgi status. Failing this kind of cytological information, it is desirable to have studies on the biological activity present in the neurohypophysis parallel to the changes in the stainable material. Such studies are almost nonexistent in fishes, but one is outstanding. Lederis (1962) showed that a high vasopressor activity is associated with those fractions of *Gadus* pituitary that contain particles resembling neurosecretory granules. In *Salmo gairdneri*, Lederis (1963, 1964) further found that on transfer to seawater there was a transitory decrease in the number of granules in some of the neurohypophysial fibers, which was correlated with a 50% reduction in the pituitary content of AVT. This suggests perhaps that this particular neurohypophysial principle is associated with the granules, although as Lederis emphasized the parallelism between AVT content and disappearance of these granules was very imperfect, only relatively few fibers displaying loss of granules, and this author emphasized that lack of stainability or of electron density are not necessarily related to absence of neurohypophysial principles.

Apart from alterations in neurosecretory material in relation to osmotic changes, the number of synaptic contacts between Type A1 fibers and pituicytes (in neurohypophysial projections into the pars intermedia) was observed to increase when *Anguilla* was transferred to seawater.

There are other observations which relate fluctuations in the amount of stainable material to the reproductive cycle. Sawyer and Pickford (1963) reported a depletion of pituitary content of isotocin, but not AVT, during the reproductive season of female *Fundulus heteroclitus*, although not in the male. Neurohypophysial factors play a role in spawning behaviour in this fish, and in studying the pituitary Sokol

(1961) found a decrease in the stainable material associated with the spawning period in both sexes. Similar observations were made in both sexes of *Tor (Barbus) tor*. A reduction in the amount of stainable material in the neurohypophysis occurs during oviposition in *Oryzias*, which may be correlated with the demonstration that injections of neurohypophysial preparations will stimulate oviposition in this fish. Olivereau (1967a) found a rarefaction of stainable material in the neurohypophysis of female eels brought to sexual maturity by treatment with a partially purified carp gonadotropin extract, and a similar change occurred in male eels brought to maturity by injections of human chorionic gonadotropin. The significance of this change is obscure, since the mechanism whereby exogenous gonadotropins activate the eel gonadotrops is unknown. Oztan (1966b) observed depletion of AF +ve material and of the ultrastructural granules from the NPO cells of *Zoarces viviparus* during the process of gonad maturation and the accumulation of these materials in the cells during early pregnancy.

Several workers have been interested in the functional relationship between the neurohypophysis and the pars intermedia, these two regions being so intimately associated in fishes. In the eel, Knowles and Vollrath (1965a,b,c, 1966a) observed that terminations of neurosecretory fibers are associated with extravascular channels bordering the pars intermedia, these channels probably draining into blood capillaries. The pars intermedia cells appear to release their products into these extravascular channels, and the authors suggested that the Type A fibers might also discharge into the channels and influence the activity of the pars intermedia cells by this route. The same Type A fibers were also observed to make synaptic contact with neighbouring pituicytes, and these synapses increased in eels briefly exposed to a white background. The pituicytes surround the tubelike extensions of the third ventricle, and in these white-exposed eels there was an increase in the PAS +ve material in these tubes. Because of these findings, it has been suggested that this PAS +ve material may be secreted by the pituicytes to act as a feedback link between the neurohypophysis and the NPO by way of the third ventricle. This is an interesting idea, but it requires more evidence before it can be considered as more than a working hypothesis; some support comes from observations in higher vertebrates, in which an association between neurons and ependymal cells, and secretion by the ependymal cells into the third ventricle, have been observed. At any rate, it is now becoming clear that pituicytes (ependymal and glial elements) may play a more important role in neurosecretory control than was hitherto supposed.

Salmo and *Perca* exhibit an anatomical relationship between Type A fibers and pars intermedia cells similar to that in the eel, but other teleosts (*Poecilia reticulata*, *Xiphophorus*, *Phoxinus*, and *Tilapia*) show nerve fibers ending directly on the pars intermedia cells, with no basement membrane or extravascular channel in between.

Apart from the eel, changes in background colour or in illumination have been found to alter the neurohypophysial structure in other teleosts. In *Zoarces*, continuous illumination in April resulted in depletion of AF +ve material and of ultrastructural granules from the cells of the NPO, together with an increase in nuclear size, compared with animals kept in the dark, although a similar experiment on September fish gave different results, with no alterations in the nuclei and accumulation of material in the cells. In *Porichthys* also, constant illumination led to a depletion of AF +ve material from the NPO. Whether these changes related to alterations in pars intermedia activity and pigmentary conditions is not known.

An isolated report suggests some relationship between the neurohypophysis and the thyroid, although this is difficult to interpret in physiological terms: Injections of TSH, or thyroxine, encouraged the movement of neurosecretory material in *Salmo* from the NPO to the neurohypophysial core.

New fields for investigation are indicated by the observations that electrical stimulation of the olfactory tract of goldfish caused depletion of stainable material from the NPO and its axons; and by the report of wide diurnal variations in the amount of AF +ve material in the goldfish NPO, variations that do not seem to be regular (rhythmic), although generally the cells stain more deeply during the day than at night. In the light of this last finding, some of the earlier work may need reconsideration, since investigators have not usually paid attention to the possibility of diurnal changes in the neurohypophysis.

There have been few observations on the Type A neurosecretory material among the cells of the pars distalis, where the function of the fibers is presumably the control of adenohypophysial activity. Knowles and Vollrath (1966b,c) described Type A3 fibers in the rostral and proximal pars distalis of eel and found that the number of these fibers, and the amount of stainable material, was greater in the silver eel than in younger animals. They suggested that this might be correlated in some way with the greater activity of the TSH cells and gonadotrops in the silver stage. In mature seawater eels, these authors observed a decrease in the AF +ve material and in Type A3 granules (140 mμ diam) in the proximal pars distalis, but fibers with smaller granules (100 mμ) were now found in this region. The significance of these changes is not known, and Knowles and Vollrath did not identify the fibers with 100 mμ granules as being distinct from the Type A3 fibers of the river eel.

"Nonstainable" fibers, probably knowles' type B

The nonstainable fibers probably originate from the nucleus lateralis tuberis (NLT) of the hypothalamus and do not stain with the classic neurosecretion stains, although they have been described as AF +ve by some workers. They can be stained with acid dyes such as azocarmine, eosin, light green, phloxin, and also with PAS Aniline blue or silver impregnation; similar staining reactions are seen in the cells of the NLT. Their pathway within the pituitary has proved difficult to follow with the light microscope, but they have been traced between cells of the eel proximal pars distalis and the trout pars intermedia, and pass toward the cells of the proximal pars distalis in *Salvelinus*.

At the electron microscope level, these nonstainable fibers appear to correspond to the Type B fibers in the eel neurohypophysis, which contain granules of less than 100 mμ diam, the granules having a central electron dense material which does not fill the bounding membrane and is less osmiophilic than the Type A granules. Similar granules are found in the NLT of eel. In *Salmo*, however, granule size in the Type B fibers is less than in the NLT, so that Follenius (1965a,b) was not certain that the NLT is the origin of the Type B fibers in this species. Moreover, in some fishes (e.g., *Phoxinus*) the NLT is said to be absent, although Type B fibers are present. It is not certain, therefore, that all Type B fibers arise from the NLT in all teleosts.

Nonstainable and Type B fibers appear to be more numerous in the anterior part of the neurohypophysial core and in the pars distalis than in the more posterior regions and the pars intermedia, although a few Type B fibers are found in the eel pars intermedia, just as some type A fibers are found in the pars distalis. Type B fibers make both neurovascular and neuroglandular contacts in both the pars intermedia and the pars distalis. In the eel, the contacts in the pars distalis are all neurovascular; that is, the neurons discharge into an extravascular space, close to the endocrine cells but separated by a distance of about 300 mμ or less. However, in *Hippocampus*, both Type A and Type B fibers were found to make direct contact with several kinds of endocrine cells in both regions of the adenohypophysis. The contacts were synaptoid for both fiber types in the pars intermedia, and Type

B fibers were shown to make synaptoid contact with three types of pars distalis cells; but satisfactory evidence of synaptoid contacts between Type A fibers and pars distalis cells could not be obtained. Direct contact between both types of fibers and pars distalis cells have also been seen in *Tilapia*, as well as terminations on the basement membrane adjacent to the endocrine cells. These findings suggest that activity of the adenohypophysial cells could be regulated by a direct dual neurosecretory innervation (Types A and B), with various kinds of contacts between the nerve fibers and the endocrine cells.

Knowles and Vollrath (1966d) found that there appeared to be an increase in the activity of Type B fibers innervating the proximal pars distalis of seawater eels in which the gonadotrops were further developed and more active than in the river eel; this is particularly interesting in that there has been evidence for some years that the NLT rather than the NPO is concerned in regulation of gonadotropic activity in teleosts.

Blood Supply to the Teleostean Pituitary

The most striking characteristic of the blood supply to the gland is that the blood reaching the adenohypophysis has passed through capillaries in the neurohypophysis; and further, there is no independent venous drainage of the neurohypophysis apart perhaps from a posterior connection to the hypophysial vein in some species. In these respects, the vascularization of the fish gland differs greatly from the tetrapod condition.

Most students of the subject have described a collection of capillaries in the neurohypophysis, forming a vascular plexus in the neurohypophysial core, close to the adenohypophysial boundary (e.g., cyprinodonts), or actually at the neuro-adeno interface. This plexus in the neurohypophysis is the primary longitudinal plexus or system of Follenius (1965a,b). In the adult *Phoxinus*, and no doubt in other species, it includes a large central longitudinal sinus. From this plexus a series of capillaries passes into the adenohypophysis and forms an elaborate network of capillaries and sinuses between the endocrine cells; Follenius (1965a,b) has termed this the "secondary centrifugal system." According to some workers, the capillaries of this system are confined to prolongations of the neurohypophysis penetrating between the endocrine cells. In contrast, in the eel, according to Knowles and Vollrath (1966a,b), these capillaries lie within a series of intervascular spaces or channels into which the neurosecretory fibers discharge and against which the endocrine cells lie, in both the pars intermedia and pars

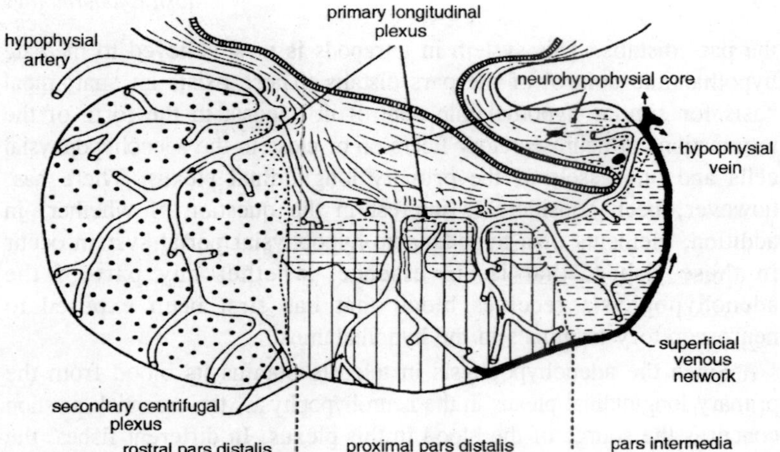

Fig. 2.3. Diagram to illustrate the main features of the blood supply to the cyprinodont pituitary.

distalis. Whatever the details, the vessels of the secondary centrifugal system obviously come into intimate association with the adenohypophysial cells and are the only source of blood for these cells.

In salmon and eel and in *Hippocampus* the size and number of capillaries in the adenohypophysis varies with age and physiological state, suggesting the possibility of a hemodynamic control of secretory activity.

From the adenohypophysial capillaries, blood is usually collected into a superficial plexus on the outer surface of the gland, and this superficial network of vessels drains posteriorly by a few veins which empty into the systemic venous system via the hypocranial or hypophysial veins.

It will be seen that this modern work indicates a centrifugal flow of blood from the primary longitudinal plexus in the neurohypophysis, contrary to the scheme proposed by Bretschnieder and de Wit (1947). Thus far, there seems to be a fair measure of uniformity among teleosts studied; however, the manner in which the primary capillary plexus receives its blood supply varies a good deal, and is very interesting because of the question of whether or not the teleostean adenohypophysis receives blood via a hypothalamo-hypophysial portal system such as is found in tetrapods, in which a primary capillary network in the median eminence region (just anterior to the pituitary stalk) is intimately associated with neurosecretory nerve endings, and blood passes from this network along the hypophysial portal veins to the capillaries in the pars distalis. This system in tetrapods is now believed to mediate hypothalamic control of the pars distalis. There exists an anatomical basis for similar hypothalamic control in teleosts in the form of the terminations of neurosecretory fibers on or close to the adenohypophysial cells and on vessels of the primary longitudinal plexus. There has, however, been considerable interest in the question of whether, in addition, elements of a hypothalamo-hypophysial portal system occur in these fishes; that is, in essence, whether any part of the adenohypophysis receives blood that has first been exposed to neurosecretory terminals in the hypothalamus.

Since the adenohypophysis in teleosts obtains its blood from the primary longitudinal plexus in the neurohypophysis, the essential question concerns the source of the blood in this plexus. In different fishes, the details vary. In most cases, a paired or single hypophysial artery (derived from the anterior carotid system) passes directly to the neurohypophysis without contributions from hypothalamic blood vessels. In *Hippocampus* and *Cichlasoma*, other branches from the internal carotid may first pass through the rostral pars distalis before entering the primary longitudinal plexus. In most cases blood from the carotid system via the hypophysial artery may be the only source of the blood in the primary capillary plexus, apart from a very few capillaries from hypothalamic vessels. In other species, the contribution from hypothalamic vessels may be more important, and it is these exceptional cases that have aroused interest in the past. In *Phoxinus*, a median ventral infundibular artery (= hypophysial artery?) supplies the ependymal floor of the infundibular recess, and just above the pituitary stalk it forms connections with a pair of ring vessels derived from the ventral hypothalamic artery. Bhargava disagrees with Barrington's description of these connections as a vascular bed, and instead he implies that they are few in number and essentially form looplike connections between the ring vessels and the ventral infundibular artery, the latter then passing into the neurohypophysial core to supply the central sinus which in the adult represents the primary longitudinal plexus. Barrington (1960) has described CAH + ve neurosecretory cells lying in the ventral hypothalamus in close association with the capillary connections ("vascular bed") and drew attention to a resemblance to the tetrapod median eminence, but Bhargava (1966) finds that these cells, which occupy the region where the NLT is found in other species, are not in fact particularly close to the capillary loops. Pending further investigations, there seems no good reason to think that *Phoxinus* exhibits any kind of rudimentary portal system.

Another slightly unusual arrangement exists in the bitterling, *Rhodeus*, in which the hypophysial artery breaks up into a superficial capillary network (rete mirabile) on the anterior face of the pituitary stalk (hypothalamic floor), from which arise the capillaries which form the primary longitudinal plexus in the neurohypophysial core. This rete mirabile is totally superficial and not associated with any special nerve endings such as characterize the tetrapod median eminence. In *Corydora* and *Cichlasoma* meningeal vessels drain into the primary longitudinal plexus, and these receive contributions from capillaries in the hypothalamus which in *Cichlasoma* are closely associated with scattered hypothalamic cells containing discrete AF +ve droplets. In *Channa punctatus*, all the blood in the pituitary is said to derive from vessels in the hypothalamus, but no details are available about possible neurosecretory associations with these hypothalamic vessels.

While these scattered observations might be interpreted as indicating some kind of portal connection between hypothalamus and adenohypophysis, the important point to emphasize is that in no teleost has any close association between hypothalamic neurosecretory nerve endings and blood capillaries been observed outside the pituitary itself; thus, whether or not these various hypothalamic capillary contributions to pituitary vascularization represent, as it were, a foreshadowing (or reminiscence) of the portal condition, there is no reason to regard them as functional teleostean portal systems. One must agree with Follenius (1965a,b) that insofar as there may be a vascular link between hypothalamic neurosecretion and adenohypophysial cells in teleosts, that link is more likely to be found in the secondary centrifugal plexus rather than in the extrapituitary vessels supplying the primary longitudinal plexus.

Hypothalamic Control of the Adenohypophysis

The limited information available about the behaviour of the teleostean adenohypophysis when deprived of direct hypothalamic connections was reviewed; from the work on pituitary transplants and *in vitro* culture, some degree of hypothalamic control of the gland was indicated, although the control seems less important than in higher vertebrates. It will be seen that anatomically there appear to be at least two routes by which hypothalamic control could be imposed on the adenohypophysis.

Neurosecretory terminations within the adenohypophysis

From ultrastructural and histochemical information, it appears that neurosecretory fibers of Type A or Type B come into more or less intimate contact with all the adenohypophysial cells. Type A apparently predominating in the pars intermedia, and Type B in the pars distalis, but with overlapping distributions in each case. Type A fibers probably originate in the NPO, and most Type B fibers probably originate in the NLT. The nature of the contact between the nerve fiber and endocrine cells ranges from immediate physical contact, synaptic in some cases, as in the pars distalis of *Hippocampus*, *Tilapia*, *Phoxinus*, and the guppy, and the pars intermedia of *Conger*, *Gadus*, *Phoxinus*, and the guppy; through cases where the nerve fiber ends on a single or double basement membrane which separates the neuron from the endocrine cells (*Tinca*, *Perca*, *Salmo*, and *Tilapia*), to the condition in which the nerve fiber ends in an extravascular space which separates the fiber terminal from the endocrine cells (*Anguilla*, *Perca*, and *Salmo*). It should be noted, however, that Knowles and Vollrath (1966a) have questioned the nature of the boundary between endocrine cells and nerve fibers that Follenius and Porte (1962) termed a "basement membrane," suggesting that it may in fact be a double membrane enclosing an extravascular space, as in *Anguilla*.

Neurosecretory terminations on capillaries in the neurohypophysial core

Apart from the neurosecretory terminations within the adenohypophysis, many workers have concluded that neurosecretory fibers make synaptic contact with the walls of the capillaries of the primary longitudinal plexus and its branches into the secondary centrifugal plexus. According to Follenius (1965a,b) it is Type B fibers that terminate in the anterior part of the primary capillary plexus, close to the pars distalis, while Type A fibers from the NPO mostly terminate on capillaries in the more posterior part of the neurohypophysial core, close to the pars intermedia. Nucleus preopticus fibers terminate on capillaries in the neurohypophysial core in *Lepidogobius*, and Honma and Tamura (1967) found that both NPO and NLT fibers terminated on neurohypophysial blood vessels in *Salvelinus*. Nucleus preopticus fibers terminate on neurohypophysial capillaries and also among proximal pars distalis cells in *Porichthys*.

A possible third route via the third ventricle

Cellular processes from cells of both the NPO and the NLT have been observed to penetrate the ependyma and project into the third ventricle, with evidence of secretion into the ventricle. Similar appearances have been described in other vertebrate groups. Stahl and Leray (1962) suggested that cells of the NLT might release their secretion into the third ventricle, to be transported thence by capillaries to the pituitary, or by being absorbed by the specialized ependymal cells lining the infundibular recess which appear to send fibers containing PAS +ve or AF +ve material toward the pituitary. In view of the evidence that NPO and NLT fibers penetrate the pituitary directly, it is difficult to assess the meaning of these appearances in terms of hypothalamic control. We should note, also, that some workers consider the specialized ependymal cells as secretory (into the third ventricle) rather than absorptive.

Thus, anatomically, the neurosecretory fibers of the neurohypophysis display features that strongly suggest they are concerned with regulation of adenohypophysial function. In addition, the neurohypophysis contains the biologically active octapeptides, AVT and IT. These or similar principles have well-documented systemic hormonal functions in tetrapods, and there is some evidence, perhaps not conclusive, that AVT and IT have peripheral actions in teleosts. How these neurohypophysial principles relate to the fiber types of the neurohypophysis is not certain, although the work of Lederis (1962, 1964) suggests that in *Gadus* and *Salmo* AVT is associated with Type A granules; this agrees with the suggestion of Knowles (1965) that Type A fibers may be concerned with elaboration of peptide molecules, whereas Type B fibers may secrete some nonpeptide material, possibly aromatic amines in line with indications from other vertebrates. The problem is to reconcile the histological and ultrastructural evidence that Type A fibers are involved in the regulation of adenohypophysial activity with the indications that the peptides which these fibers probably secrete may have peripheral activities. At the present time too little evidence is available to permit further analysis of this problem. More work is needed, particularly to demonstrate definitively the functional significance of the neurohypophysial invasion of the pars distalis, so characteristic of teleosts. Direct innervation of the pars intermedia is found throughout the vertebrates, but the anterior neurohypophysis-pars distalis association is a teleostean speciality, and it may be that the extensive neurosecretory innervation of the pars distalis cells in this group evolved as an adjunct or supplement to the capillaries of the secondary centrifugal plexus. It is worth recalling the comparison, frequently made, between the median eminence of other gnathostomes and the anterior neurohypophysial core of teleosts, a comparison all the more apt in the light of recent information, both being regions where hypothalamic neurosecretory fibers terminate on capillaries which convey blood to the pars distalis; it may be that the anterior capillaries of the secondary centrifugal plexus are functionally equivalent to the hypophysial portal vessels, conveying products of neurosecretory cells (mainly Type B) to the pars distalis, the median eminence and portal plexus being as it were enclosed within the pars distalis.

Comparisons between the tetrapod and fish pituitary regions are complicated by the presence in nearly all fishes of the saccus vasculosus, a thin-walled folded sac growing out from the third ventricle (infundibular recess) posterior to the pituitary. This structure, despite its close proximity to the pituitary (particularly marked in *Acipenser* and elasmobranchs) seems to have no functional connection with the gland. In the vertebrate embryo there arises from the postero-ventral floor of the hypothalamus a posteriorly directed downgrowth, the saccus infundibuli. In fishes, this gives rise to the saccus vasculosus, and the fish neurohypophysis arises more anteriorly from an area of the hypothalamic floor that in tetrapods and lungfishes gives rise to the median eminence. In tetrapods, the saccus infundibuli gives rise to the pars nervosa, the main part of the neurohypophysis. Thus embryologically, the neurohypophysial core of fishes is not comparable to the tetrapod pars nervosa, which, is more closely comparable to the fish saccus vasculosus; but in functional terms the pars nervosa may be likened to the posterior part of the fish neurohypophysis, particularly considering the extensive innervation of the pars intermedia by neurohypophysial fibers in many tetrapods. On both embryological and functional grounds, the tetrapod median eminence may be comparable to the anterior region of the teleostean neurohypophysial core, and to the shallow anterior part of the neurohypophysis dorsal to the pars distalis (the infundibular floor) in the primitive bony fishes.

Pituitary Gland in Primitive Bony Fishes

Bony fishes belong to the class Actinopterygii, and the vast majority belong to the most recently evolved group of actinopterygians, the superorder Teleostei. In addition, there are a few surviving members of the more ancient superorders Chondrostei and Holostei. The pituitary gland in these primitive forms is of obvious interest in possibly indicating the evolutionary developments which lead to the extreme specializations exhibited by the teleostean gland. Unfortunately, published information is almost entirely limited to anatomical and histological descriptions, with little experimental data to help identify the various cell types.

Superorder Chondrostei

Order Palaeoniscoidei (Polypterus and Calamoichthys)

General accounts of the pituitary in these fishes have been given recently by Dodd and Kerr (1963) and Wingstrand (1966a), and Kerr

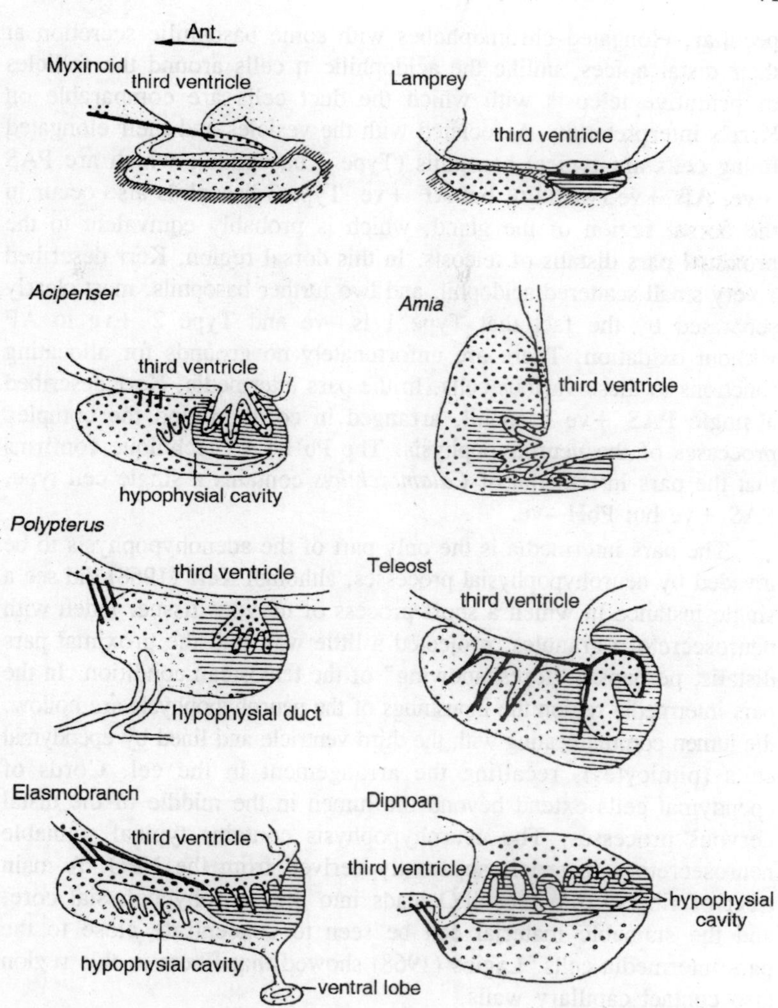

Fig. 2.4. Diagrammatic midsagittal sections to illustrate the main features of pituitary structure in the different fish groups.

(1968) has made a detailed histological study. The most remarkable feature is the presence in the adult of a persistent remnant of the ventral part of Rathke's pouch, the hypophysial duct, which opens ventrally into the roof of the mouth and dorsally into an expanded space in the ventral part of the pars distalis. From this space, short diverticuli penetrate the pars distalis, forming vesicles in sections. Kerr (1968) equates this ventral part of the gland with the rostral pars distalis of teleosts, although the cells lining the vesicles and duct are peculiar, elongated chromophobes with some basophilic secretion at their distal apices, unlike the acidophilic η cells around the follicles of primitive teleosts with which the duct cells are comparable on Kerr's interpretation. Associated with the vesicles and their elongated lining cells are typical basophils (Type 3 basophils), which are PAS +ve, AB +ve (weakly), but AF +ve. Type 3 basophils also occur in the dorsal region of the gland, which is probably equivalent to the proximal pars distalis of teleosts. In this dorsal region, Kerr described a very small scattered acidophil, and two further basophils, most clearly separated by the fact that Type 1 is –ve and Type 2 +ve to AF without oxidation. There are unfortunately no grounds for allocating functions to these various cells. In the pars intermedia, Kerr described a single PAS +ve basophil, arranged in cords around the complex processes of the neurohypophysis. The PbH-PAS technique confirms that the pars intermedia of *Calamoichthys* contains a single cell type, PAS +ve but PbH –ve.

The pars intermedia is the only part of the adenohypophysis to be invaded by neurohypophysial processes, although Kerr (1968) did see a single instance in which a short process of nervous tissue, laden with neurosecretory granules, projected a little way into the proximal pars distalis, perhaps a "foreshadowing" of the teleostean condition. In the pars intermedia region the inpushings of the neurohypophysis are hollow, the lumen communicating with the third ventricle and lined by ependymal cells (pituicytes), recalling the arrangement in the eel. Cords of ependymal cells extend beyond the lumen in the middle of the distal nervous processes. The neurohypophysis contains typical stainable neurosecretory material, apparently derived from the NPO. A main tract of fibers from the NPO leads into the neurohypophysial core, and the stainable material can be seen to concentrate close to the pars intermedia cells. Lagios (1968) showed that fibers in this region also contact capillary walls.

A subsidiary tract from the NPO lies ventral to the main tract in the hypothalamic floor and shows prominent stainable material just anterior to the tip of the pituitary. In this region the neurosecretory fibers are closely mingled with a plexus of blood capillaries, and branches from this plexus pass into the pars distalis across a connective tissue pad at the front end of the gland. As Kerr (1968) pointed out, there is little reason to doubt that here we have a typical median eminence and extrapituitary portal system. This is confirmed by ultrastructural studies, which showed that in the median eminence of *Calamoichthys* there are neurosecretory fibers with granules 100-150 mμ diam which terminate on extravascular spaces around capillaries in a manner indistinguishable from that in the amphibian median eminence. According to Lagios, the portal vessels contribute the sole vascular supply to the pituitary of *Calamoichthys*, but Kerr (1968) thought that part of the rich vascular plexus in the neurohypophysial tracts in the pars intermedia of *Polypterus* and *Calamoichthys* derived from vessels in the wall of the brain.

In addition to the portal vessels in the connective tissue pad, the connective tissue sheet between the pars distalis and the infundibular floor (= anterior part of neurohypophysis) is highly vascular, and its capillaries are in close proximity to the neurosecretory filters in the infundibular floor. It is possible that some neurovascular exchange may occur here, as appears to be the ease in holosteans, capillary branches also penetrating the pars distalis from the connective tissue plexus.

Order Acipenseroidei (Acipenser and Polydon)

The sturgeon pituitary has been studied many times, particularly by Russian workers, but there is still little detailed information about its histophysiology. The most notable anatomical feature is the persistence of a large central hypophysial cavity or cleft, apparently a remnant of the dorsal part of Rathke's pouch, although unlike the palaeoniscids the sturgeons do not retain a hypophysial duct to the roof of the mouth. The hypophysial cleft largely separates the posterior pars intermedia from the anterior pars distalis, and also largely splits the latter into dorsal and ventral portions. The cleft gives off numerous tubular extensions, particularly dorsally but also anteriorly and ventrally. In addition to these tubules there are in the anterior and ventral pars distalis closed vesicles, with no connection to the cleft. In the absence of detailed embryological information their morphological status is uncertain, but Kerr (1949) assumes they are derived from the cleft tubules. These vesicles are presumably homologous with the follicles of the rostral pars distalis of holosteans and primitive teleosts. *Acipenser fluvescens* shows acidophils, often elongated, oriented around these vesicles, together with equally abundant rounded basophils and chromophobes. However, in the rostral region of *A. stellatus*, where the vesicles predominate, Barannikova (1949) described strongly acidophilic cells around the vesicles, together with amphiphils and only a few basophils. Both authors agree that the vesicles usually contain basophilic colloid as in holosteans and teleosts.

The ventral pars distalis, beneath the hypophysial cleft, was considered by Kerr to be an extension of the rostral pars distalis, with the same cellular composition, but Barannikova distinguishes this zone from the rostral region, since its acidophil cells stain more weakly and it contains many more basophils and small chromophobes. In the dorsal pars distalis, above the cleft, Kerr described regular columns of cells, arranged around tubular extensions of the cleft, containing acidophils and basophils. Barannikova (1949) states that this region contains the faint acidophils, basophils, and chromophobes also found in the ventral zone, together with strongly staining acidophils. By bioassay of dissected glands of *A. stellatus*, Barannikova (1949) located gonadotropin secretion in the ventral and dorsal zones and showed that the activity was stronger in the ventral zone; the basophils of these two zones displayed secretory changes correlated with gonadal development, identifying these cells as gonadotrops. The gonadotrops display marked development of acidophilic granules during their activity cycle, reminiscent of the R granules of *Poecilia* and other teleosts.

Since the gonadotrops are associated in both dorsal and ventral zones with a weakly staining acidophil, apparently distinct from the rostral acidophils, we may provisionally conclude that these two zones in *A. stellatus* correspond to the teleostean proximal pars distalis. However, as in holosteans, there is no very sharp separation of the two regions of the pars distalis, with broad mixed zones exhibiting all the cell types.

The pars intermedia is very large, and consists almost entirely of large basophils together with small numbers of chromophobes and acidophils. The neurohypophysis is separated from the adenohypophysis by a connective tissue layer, except in places over the pars intermedia where long hollow neurohypophysial processes penetrate deeply among the endocrine cells, with ependymal cells lining the cavities of the processes in the usual arrangement. No processes penetrate the pars distalis. Neurosecretory axons from the NPO and possibly also from the NLT terminate on blood vessels in the neurohypophysial processes. The connective tissue sheet between the infundibular floor (anterior neurohypophysis) and pars distalis is vascularized, and as in holosteans neurosecretory fibers from the NPO may terminate in relation to these capillaries, forming a kind of median eminence.

Superorder Holostei (Amia and Lepisosteus)

An account of the anatomy and histology of the pituitary in both genera is available, and one of us has investigated the glands of 4 *Amia* and 3 *Lepisosteus* using the modern techniques.

Neither genus preserves a hypophysial cavity or duct in the adult. The gland lies attached along most of its length to the infundibular floor (anterior neurohypophysis) and in *Amia* is relatively shorter and deeper than in *Lepisosteus*. Behind the pituitary is a large saccus vasculosus. Both genera display to a marked degree an intermingling of all the cell types in the pars distalis, as in the sturgeons, so that it is difficult to divide the area clearly into rostral and proximal regions.

In both genera, the rostral tip of the gland consists of the characteristic closed vesicles or follicles, containing basophilic material (PAS +ve, AB +ve, and AF +ve), and it seems possible from embryology that these do indeed represent diverticuli from the hypophysial cleft. In *Amia*, the follicles are lined by elongated acidophils, which are erythrosinophilic in the Aliz B technique and resemble the η cells of the eel (*acidophil 1*). Also in the follicle walls, but not usually reaching the lumen is a second cell type, *acidophil 2*, again erythrosinophilic but also taking Alizarin blue, so that it usually stains a blue-red in contrast to the scarlet of acidophil 1; the acidophil 2 is strongly PbH +ve, and it may be the corticotrop. A few basophils and acidophils from the main (proximal) pars distalis may be mixed with the follicles, especially ventrally. Also ventrally, the follicles are joined by a very peculiar cell type. This is strongly PAS +ve, Af +ve, AB +ve, and PbH +ve, and this is termed *basophil 1*; however, after Aliz B, its coarse retractile granulation is brilliantly stained with erythrosin. In addition, there are usually present a few large Aniline blue +ve granules which, together with the large size of the granules and difference in cell shape, distinguish basophil 1 in this technique from acidophil 1. A mantle of basophil 1 cells extends laterally and ventrally round the gland, enclosing the rest of the pars distalis and the pars intermedia, containing an admixture of acidophils 1 and 3.

The proximal region of the pars distalis is composed of dorsoventrally oriented cords of cells. They include an *acidophil 3*, sharply separated from acidophils 1 and 2 by staining with orange G after Aliz B, and possibly an α cell (like the teleostean α cell, it has a slight PAS affinity); this acidophil is most numerous dorsally and centrally, by the connective tissue interface with the infundibular floor (anterior neurohypophysis), but it occurs throughout the region. Two types of basophil are present in the proximal pars distalis cords. *Basophil 2* is smaller, with rounded contours, stains a clear blue in Aliz B, is slate blue in Ox-AB-PAS-OG, is AF +ve and PbH +ve;

this cell type occurs mostly in the dorsal and central parts of the region. *Basophil 3* is larger, with angular contours, and stains lavendar in Aliz B (it contains scattered red granules as well as dull blue granules), with a few large blue granules close to the nucleus in many cases. It stains magenta with Ox-AB-PAS-OG, and is more strongly AF +ve and PbH +ve than basophil 2. It occurs mainly in the ventral and lateral region. These basophils 2 and 3 probably correspond to the two basophils described by Kerr.

The pars intermedia is elaborately invaded by branching processes of the neurohypophysis, which are hollow (at least proximally), with a lining of ependymal cells. Two cell types are distinguishable in the pars intermedia cell cords: a predominant PbH +ve cell, with the club shape often found in its teleostean homolog, the prolongation contacting the neurohypophysial interface; and a rather scarce PAS +ve cell, which tends to be rounded. The PbH +ve cell is blue in Aliz B preparations but of widely varying shades. The PAS +ve cell is amphiphilic, its colouration varying from red to mauve after Aliz B.

The neurohypophysis is essentially the thin floor of the infundibulum, separated from the adenohypophysis by a connective tissue sheet, except where it deepens and penetrates the pars intermedia as the neurohypophysial core. The penetration is by a hollow downfolding of the infundibular floor, forming a large infundibular funnel which penetrates deeply to the middle of the neurohypophysis. No processes appear to penetrate the pars distalis. The neurohypophysial core is rich in neurosecretory material, AF +ve, AB +ve, PAS +ve, PbH +ve. Aniline blue +ve, and CAH +ve. Sathyanesan and Chavin (1967) have traced fibers with these staining properties from the NPO into the neurohypophysial core, and observed some of these fibers terminating on blood vessels in this region in *Lepisosteus*; in this genus, they also found fibers from the NLT, AF –ve, but CAH +ve, directed toward the pituitary.

A simple portal system appears to be present; whether or not it is extrapituitary depends on how one interprets the status of the infundibular floor above the pars distalis. In this region, the connective tissue sheet separating the nervous tissue from the pars distalis is extremely vascular, with a plexus of capillaries and sinuses. From this plexus branches pass up into the neural tissues and down to vascularize the pars distalis. Quantities of AF +ve neurosecretory material occur close to these capillaries in the neural tissue just above the posterior part of the pars distalis, suggesting that NPO fibers may terminate on capillaries of the mantle plexus and their product may be transported into the pars distalis. In addition, in the more anterior part of the infundibular floor, above the anterior proximal pars distalis and the rostral zone, Herring bodies and neurosecretory grains of "nonstainable" type lie close to the mantle plexus capillaries, presumably deriving from the NLT, which displays similar material in its neurons.

Thus, in all the primitive bony fishes, a portal system exists in the form of capillaries linking the infundibular floor, rich in neurosecretory terminations (= median eminence) and the pars distalis capillary network. It is easy to conceive how the teleostean condition could be derived, if this infundibular floor and its mantle plexus were to be folded into the pars distalis, resulting in an enclosed median eminence (anterior neurohypophysial core) with its primary longitudinal plexus.

The pituitary of *Lepisosteus*, although of different shape, resembles the *Amia* gland in essentials. Main points of difference are: the pars intermedia is larger, with wider neural interdigitations and a small infundibular funnel; the rostral follicles are smaller, and intermingle more freely with proximal pars distalis cells. *Lepisosteus* displays acidophils 1, 2, and 3, and the mantle of basophils 1 seen in *Amia*; however, basophils 2 and 3 are less easily separable, apart from a much sparser granulation in basophil 2 and its more central location. The neurohypophysis and infundibular floor are essentially as in *Amia*.

In both genera, the mantle plexus in the infundibular floor appears to be supplied by arteries which enter the infundibular region anteriorly between the optic nerves and the pituitary, although additional sources may be present.

Pituitary Gland in Lungfishes

As surviving, though distant, relatives of the animals ancestral to the tetrapods, the pituitary in lungfishes (Dipnoi) is of great interest, but there have been few studies on the gland using modern cytological techniques. The recent work of Wingstrand (1956), Kerr and van Oordt (1966), and van Oordt and Kerr (1966) all describes the pituitary in the African lungfish, *Protopterus* sp.

The organization of the pituitary resembles that of amphibians rather than other fishes. The adenohypophysis includes a distinct pars intermedia, dorsally placed and intimately associated with the neurohypophysis, and a ventral pars distal is, separated from the intermedia by the hypophysial cleft. In the pars distalis, as in amphibians, the cell types are intermingled and not grouped into zones as in so many fishes.

Five cell types can be distinguished in the pars distalis. Although the lack of direct experimentation makes it impossible to associate each of these firmly with the secretion of a particular hormone, in their staining reactions and distribution these cells resemble the cells of the amphibian pars distalis for which there is adequate experimental backing for functional identification. On the basis of these similarities, and considering the time at which the dipnoan cell types appear during ontogeny, Kerr and van Oordt (1966) have made tentative functional identification of some of these cells.

Two acidophils can be identified. Type 1 acidophil is erythrosinophilic after Aliz B and is distributed throughout the gland. Type 2 is orangeophilic and is confined to the posterior region. Presumably these cells secrete prolactin and growth hormone. Three types of basophils occur. Type 1 and Type 2 both have the typical staining reactions of basophils (PAS +ve, AB +ve, AF +ve, Aniline blue +ve), but can be separated by details of shape, distribution, and granulation. Type 3 basophils are restricted to the anterior tip of the gland and are violet after Aliz B, PAS +ve but AF –ve and AB –ve. The Type 1 basophil appears very early in development and may be a thyrotrop. The Type 3 basophil appears later in ontogeny, and strongly resembles the amphibian LH cell in location and staining properties. The Type 2 basophils are found only in adult fish, in which they may be very abundant, and are possibly FSH cells. As far as one can compare the descriptions, Godet (1964) suggested a similar functional identification on the basis of the tinctorial properties of the cells. He observed that all three basophils regress during estivation, and correlated this with regression of the thyroid, cessation of spermatogenesis, and atrophy of sex accessories.

The pars intermedia and neurohypophysis are closely associated to form a dorsal neurointermediate lobe, separated from the pars distalis by the neurohypophysial cleft. *Protopterus* has a thick intermedia, composed of hollow tubules interdigitating with the neurohypophysial processes. The cavities of the intermedia tubules often communicate with the hypophysial cleft, from which they are probably ultimately derived, but in the adult many are closed. In *Neoceratodus* the hypophysial cleft also sends diverticuli into the pars distalis, as in *Acipenser*, and in *Lepidosiren* the pars intermedia is just a thin layer of cells. Two pars intermedia cells occur in *Protopterus*: one is weakly

PAS +ve and AB +ve but strongly AF +ve, while the other is strongly PAS +ve but AB –ve. The latter cell type is more abundant in younger fish. There is some indirect evidence that MSH is secreted by this region as in other fishes. Total hypophysectomy leads to melanin concentration, but removal of the pars distalis alone causes melanin dispersion together with hypertrophy of the pars intermedia. This hypertrophy together with persistent skin darkening suggests an enhanced secretion of MSH, and it is possible that as in many other vertebrates the pars intermedia is innervated by inhibitory hypothalamic nerve fibers which are damaged when the pars distalis is removed.

Adaptation to aerial life in the cocoon during estivation is accompanied by melanin dispersion in *P. aethiopicus*, and by degranulation of the pars intermedia cells, although no histological details were given. Under these conditions, the cavities of the intermediate lobe tubules enlarge.

The dipnoan neurohypophysis is comparable embryologically to the pars nervosa of tetrapods rather than the neurohypophysis of other fishes. It consists basically of a system of hollow tubules of nerve fibers, shown in development to arise as outpushings of the infundibular cavity, but in the adult this relationship is lost and the tubules mostly become solid. Ependymal cells, originally surrounding the tubule cavities, extend in the nervous processes together with abundant AF +ve neurosecretory material. In the adult, the intertwining of the nervous processes with the cords of the pars intermedia are very complex. The neural processes contain more stainable material in aquatic *Protopterus* than in aerial (estivating) fish, and the amount of this AF +ve material decreases in fish transferred from water to aerial cocoons. When air is blown into the mouths of aquatic fish, there is a marked increase in the amount of neurosecretory material after 3 hr. It is not clear in all these experimental conditions whether an increase in the amount of material should be taken to mean increased synthesis or decreased release of the neurohypophysial principles.

Arginine vasotocin has been identified in the pituitary of *Protopterus*, and it has been shown to have diuretic and natriuretic effects when injected.

Anteriorly, the interdigitating neurohypophysis connects to the thin floor of the infundibulum which, as in the primitive bony fishes, is separated from the pars distalis by vascular connective tissue. This region passes at the anterior margin of the gland into a distinct median eminence, comparable to that of urodele amphibians. Its capillary plexus is supplied by fibers from the NPO, and portal vessels pass from this plexus to supply the pars distalis.

Pituitary Gland in Elasmobranchs

The adenohypophysis in this group is at first sight very unlike that in other fishes. It is divided into a very large pars intermedia, elaborately entwined with the neurohypophysis to form a neurointermediate lobe; and an elongated pars distalis, composed of a dorsal lobe (rostral lobe or anterior lobe) extending forward close beneath the infundibular floor, and a ventral lobe, attached to the dorsal lobe by a stalk and associated with the floor of the cranium. In sharks and certain rays, the entire pars distalis is hollow, the anterior part of the dorsal lobe (head of dorsal lobe) containing vesicles or tubules communicating with the central hypophysial cavity, and the posterior part (tail of dorsal lobe) consisting simply of folds of tissue around the hypophysial cavity. In other skates and rays, the hypophysial cavity may be small or obliterated, and the dorsal lobe is a compact mass of cords and clusters of cells. The ventral lobe in all cases is hollow, described as containing vesicles. In the aberrant chimaeroids, the gland displays a hollow dorsal lobe, and a separated structure, composed of follicles of cells, which may represent a detached ventral lobe. All these spaces and vesicles may contain a colloid, which is PAS +ve, AF +ve, and AB +ve. The suggestion has been made that this colloid in the vesicles of the ventral lobe may represent a store of gonadotropins and thyrotropin, but Mellinger (1962b) believes that secretion of the colloid throughout these hypophysial spaces is a nonspecific function of the cells lining the cavities, including the endocrine cells and certain nonendocrine cells. One is reminded of the follicles of the rostral pars distalis in some actinopterygians, which frequently contain a PAS +ve colloidal material.

Experiments involving surgical hypophysectomy (total or partial) and replacement therapy with mammalian hormones have suggested that the ventral lobe secretes gonadotropin(s) in dogfish and skate. Assays of different regions of the gland for TSH also locate thyrotropic function in the ventral lobe. ACTH activity has been located in the head of the dorsal lobe by bioassay.

The limited information available about cell types in the gland does not always correlate with this physiological data. In the head of the dorsal lobe are found strongly PAS +ve cells. Mellinger (1962b, 1966) terms these the "T cells," and showed in various species that they are AF –ve and AB –ve and acidophilic. The T cells he equates with the β cells of Della Corte and Chieffi (1961) and Chieffi (1962); the Italian workers thought that these cells produced TSH, but Mellinger believes that they secrete growth hormone. The T cells occupy the region in which deRoos and deRoos (1967) located ACTH activity. In the tail of the dorsal lobe, Mellinger finds Q cells, PAS +ve to a variable extent, AF –ve and AB –ve and acidophilic, which he suggests secrete ACTH. Some of these possibly correspond to the Italian workers' α cells, which they suggested might secrete growth hormone, and some of their δ cells, which they thought might produce prolactin. In the ventral lobe, there is more agreement about attribution of functions. Mellinger describes PAS +ve, AF +ve, and AB +ve cells in this lobe, which he divides on the size of their granules into V cells (gonadotrops) and X cells (thyrotrops); the Italian workers described here a single basophil which they thought secreted FSH.

In the chimaeroid, *Hydrolagus*, there are in the head of the dorsal lobe acidophils, AF +ve cells and chromophobes, while the tail of this region contains acidophils, basophils, and chromophobes. The putative ventral lobe displays small cells containing PAS + ve granules, an AF +ve cell type, also with PAS +ve granules, and chromophobes.

While the functional identity of the various elasmobranch cells is far from established, taking the physiological localizations of functions together with the histological data, it is clear that in the sharks and rays the dorsal and ventral lobes, taken together, do indeed have the functions of the pars distalis, as suggested in the nomenclature we have adopted. Alternative schemes, in which the dorsal lobe is termed the "anterior lobe" or the "rostral and proximal pars distalis", or simply "rostral lobe" all imply in one way or another potentially misleading comparisons with parts of the gland in tetrapods and teleosts.

The pars intermedia lies below the thin neurohypophysial layer and is penetrated by neurohypophysial fibers to a variable extent, forming a neurointermedia lobe. The intermedia cells may be arranged in distinct lobules separated by highly vascular connective tissue, or the cell cords may fuse to form a mass of cells with an irregular plexus of blood vessels. The penetration of nerve fibers between the intermedia endocrine cells may be slight as in *Squalus* and *Etmopterus* or very extensive as in *Scylliorhinus Torpedo*, and *Raia*. Most workers agree that only one cell type occurs in the pars intermedia, weakly PAS +ve and acidophilic. Knowles (1965) differentiated peripheral and central cells, on the basis of shape and ultrastructure, in *Scylliorhinus*, and suggested they may have different functions. In the peripheral cells, he distinguished a synthetic region at the apex of the cell and a hormone release region at the opposite pole close to a blood vessel. Knowles (1965) described Type A neurosecretory fibers from the NPO terminating on the synthetic pole of the peripheral cells, and Type B fibers, possibly from the NLT, or arising within the neurohypophysis, terminating on the release pole of the cell. Knowles interpreted these findings as indicating that hormone synthesis and hormone release are under independent neurosecretory control. Other workers have observed terminations of Type A fibers on the pars intermedia cells and on blood vessels in this region, and Meurling (1963) described AF –ve fibers, possibly equivalent to Knowles' Type B, ending on pars intermedia cells and blood vessels in *Etmopterus*. However, Mellinger (1962a) described only Type A fibers in the neurointermediate lobe of *Scylliorhinus*, and further showed that destruction of the NPO- hypophysial tract led to activation of the pars intermedia cells and release of excessive MSH, indicating that the lesion had removed inhibitory control of both synthesis and release. Chevins (1968) found in *Raia* that not only tract section and ectopic transplantation of the neurointermediate lobe but also destruction of the NPO itself led to excessive MSH secretion, which suggests either a single inhibitory control of both synthesis and release, contrary to Knowles' hypothesis, or that both Type A and Type B neurons originate in the NPO, which is contrary to the histological and ultrastructural evidence.

Large osmiophilic and acidophilic globules occur in the elasmobranch pars intermedia, as in other vertebrates, and have been interpreted as the products of cellular degeneration or as a hormone store.

As in other groups, the pars intermedia has been shown to secrete MSH, the evidence coming from surgical removal of the neurointermediate lobe, bioassay of different pituitary regions, and lesioning of the NPO-neurohypophysial tract, which caused melanin dispersion associated with hyperactivity of the pars intermedia cells. Apart from this last study, there have been virtually no observations on natural or experimentally induced changes in the pars intermedia cells other than reports of the appearance of giant cells in this region in female *Scylliorhinus* and *Torpedo* in relation to the sexual cycle. Knowles (1965) did not observe these cells in his work on *Scylliorhinus*, and their significance is unknown.

As in other fishes, the neurohypophysis develops from the thin floor of the infundibulum, dorsal to the pars intermedia. Above the pars distalis, the infundibular floor remains thin, similar to that in primitive bony fishes. The neurohypophysis may be very small (e.g., *Chimaera*), or together with the intermedia it may form two large lateral lobes, as in *Scylliorhinus*. Neurosecretory fibers, AF +ve and AF –ve, myelinated and nonmyelinated, penetrate downward to form a relatively thin nervous layer above the intermedia, and also penetrate among the intermedia endocrine cells to some extent. In older specimens of *Squalus*, Meurling (1962) described diverticuli from the infundibular cavity lined with ependymal cells, which penetrate with nervous processes into the pars intermedia, an arrangement similar to that in the eel and primitive actinopterygians. The more dorsal region of the neurohypophysis is often poor in stainable neurosecretory material, the greatest accumulation of which is in the region immediately dorsal to the pars intermedia. Some species display a distinct membrane between the neural tissue and the intermedia cells, containing a network of blood capillaries; in these cases there is little penetration of nerve fibers between the endocrine cells. In other species, the membrane has largely disappeared, and the fibers may penetrate between the cell cords either in broad bundles (e.g., *Scylliorhinus*) or irregularly (e.g., *Raia*, *Torpedo*, and *Pristiurus*).

Knowles' demonstration (1965) of Type A and Type B fibers in the dogfish neurointermediate lobe has been discussed above. Meurling (1963) also traced AF –ve fibers, presumably corresponding to Type B, to terminations on pars intermedia cells. The origin of Type B or AF –ve fibers is uncertain, although some at least probably come from the NLT. The AF +ve (= Type A?) fibers have been traced back to the NPO. The NPO Type A fibers have been traced into the neurointermediate lobe to terminations on pericapillary spaces, pituicytes, and gland cells; and other Type A fibers pass to the ventral hypothalamus (anterior infundibular floor) just above the dorsal lobe of the pars distalis, where they become associated with a capillary network to form a median eminence. Type B fibers have also been traced to the median eminence. The existence of these two fiber types has been established in the hypothalamo-neurohypophysial tract and neurointermediate lobe of *Raia*, in addition to the dogfish, *Scylliorhinus*.

There are no published accounts of induced changes in the amount of material in the neurohypophysis of these fishes. Peptides with the usual properties occur in the neurohypophysis of elasmobranchs, although apparently differing from the principles in other groups. Perks and Dodd (1960) showed that after section of the hypothalamo-neurohypophysial tract of *Scylliorhinus* the oxytocic activity of the neurointermediate lobe eventually disappeared, and Chevins (1968) found that typical AF +ve neurosecretory material eventually disappeared from the tract and neurointermediate lobe of *Raia* following ablation of the NPO.

The blood supply to the elasmobranch pituitary has attracted great interest and is reviewed by Meurling (1967a). The vascular supply in chimaeroids is described by Sathyanesan (1965b), Jasinski and Gorbman (1966), and Meurling (1967b). Apart from the dorsal pars distalis in the chimaeroid, *Hydrolagus*, it appears that each lobe of the pituitary receives arterial blood directly from either the vertebral or internal carotid arteries. In addition, there is evidence for a hypophysial portal system, although its details differ in different accounts. All authors agree that there is a region in the anterior infundibular floor (anterior hypothalamus) where neurosecretory axons from the NPO (and some from the NLT) are grouped around a capillary plexus in a way that suggests a neurohemal organ. Ultrastructural studies confirmed the presence in this region of neurosecretory terminations on the capillary walls, strongly resembling those in the mammalian median eminence. From this primary plexus in the median eminence, most of the blood passes backward in capillaries, some of which supply the tail of the dorsal lobe, while a few or many pass back to supply the neurointermediate lobe. The head of the dorsal lobe is said by some authors to receive portal blood from the anterior part of the median eminence, but other workers have not described this. Most workers have said that the ventral lobe receives no blood from the portal vessels, but in *Raia* sp. Chevins (1968) finds that blood enters the ventral lobe from sinuses in the dorsal lobe, so that at least some of this blood must be portal in origin.

In the chimaeroid, *Hydrolagus*, the portal system is similar. The dorsal lobe, however, does not receive any direct arterial blood, and numerous portal vessels from the elongated median eminence supply all parts of the dorsal lobe. No portal vessels enter the neurointermediate lobe in this fish, in contrast to sharks and rays, and the putative ventral lobe receives no portal blood at all. As these authors point out, this may well be the most primitive of all vertebrate hypophysial portal systems.

All authors agree on the absence of direct neurosecretory innervation of the pars distalis, in contrast to teleosts, so that the portal system would seem to be the only route for hypothalamic control.

The existence of a portal supply to the neurointermediate lobe is puzzling in view of the rich neurosecretory innervation of this region, and the functional significance of this part of the portal system awaits future work. The portal supply to the pars distalis is clearly similar to that in primitive actinopterygians and dipnoans, and as in those groups is to be correlated with the absence of direct innervation of the pars distalis.

Pituitary Gland in Cyclostomes

In this group, we find that the most primitive of all pituitary glands has a much simpler structure than that of gnathostomes. Both neural and glandular components can be recognized, but direct or vascular communication between the two is very limited. Many of the cells in the adenohypophysis appear chromophobic, especially in young animals, which hampers investigation of the histophysiology. Another great difficulty in studies on these animals is that their primitive evolutionary position makes it hazardous to assume that all the hormones secreted by the gnathostome pituitary are also produced by the cyclostome gland. In fact, evidence for the secretion of hormones other than gonadotropin(s), MSH, and AVT in lampreys is very insecurely based.

The pituitary in myxinoids (hagfishes) appears to be more primitive than that of lampreys (Petromyzontidae). It consists of follicles and clusters of cells embedded in connective tissue below the hypothalamus, and there is no clear cytological differentiation between pars distalis and pars intermedia. Some workers report that the connective tissue septum between neurohypophysis and adenohypophysis may be missing posteriorly, so that the two components here are in contact, but this has not been observed by other workers. In the anterior region of the adenohypophysis, most of the cells are chromophobic, but a few basophils and acidophils have been recognized. Two types of basophils have been differentiated, both PAS +ve: one is weakly AB +ve and is grouped in follicles sometimes associated with accumulation of intercellular PAS +ve colloid, and with cytoplasm rich in SS/SH groups. The second basophil stains with AF, forms signet ring cells after gonadectomy, and may be the gonadotrop. Two erythrosinophil cell types are present, one with fine granulation which responds to adrenocortical inhibitors, cortisol and thiourea, which may be the ACTH cell; and a second with coarser granulation is activated by reserpine treatment, and may secrete prolactin. The predominant cell type in the caudal region is PAS +ve and is presumed to be the source of

MSH, although there is no evidence for the secretion of this hormone in myxinoids.

The gland in lampreys is organized in a more familiar pattern. The pars distalis is embedded in connective tissue, which separates it from pars intermedia and neurohypophysis; this region is divisible into rostral and proximal parts by analogy with the teleost gland. The posterior pars intermedia is separated from the neurohypophysis only by a vascular plexus.

In the rostral pars distalis are found chromophobes, and basophils which arc PAS +ve, AF +ve, and contain SS/SH groups. In various lampreys, these basophils increase in number and staining affinities at metamorphosis, when the entire pars distalis may increase in size. Following metamorphosis these cells exhibit changes which are not easy to interpret, but which have been taken to indicate a gonadotropic function; but these changes have not always been observed. The electron microscope shows that there are two kinds of basophils in this region, differentiated by granules size.

The proximal pars distalis exhibits acidophils and basophils, but the majority of cells are chromophobic and appear most active at metamorphosis in *Lampetra planeri* or during the anadromous migration in *L. fluviatilis*. Changes in the acidophils have been described, but are impossible to interpret. The basophils are PAS +ve and AF +ve, and they are not numerous. They have been described as most active during metamorphosis or in the spawning migration and as reduced in activity after spawning. Roth (1957) and Evennett (1963) observed that these PAS +ve cells in the proximal pars distalis became increasingly chromophilic during the gonadal maturation, a change that could be prevented by gonadectomy. Since partial hypophysectomy indicated that gonadotropic function located in the proximal pars distalis, Evennett (1963) concluded that these PAS +ve cells are gonadotrops. On the other hand, Larsen (1965), working on the same species as Evennett (*L. fluviatilis*), found that both the rostral and the proximal regions secreted gonadotropins; thus, the basophils in both regions could be gonadotrops. The changes described in the rostral basophils would fit this idea, although their activation during metamorphosis could indicate a thyrotropic function, the thyroid developing at this time from the larval endostyle. At the ultrastructural level, Bage (1967) differentiated three types of chromophils as well as chromophobes in the proximal region.

The pars intermedia has been shown to secrete MSH. There is no significant penetration of this region by the neurohypophysis. Many of the intermedia cells are chromophobic, but some authors have recognized a single chromophil, PAS +ve and azocarmine +ve (i.e., amphiphilic). Others have described two chromophils: one AF +ve and PAS –ve and elongated toward the neurohypophysis although situated ventrally, the other AF +ve and PAS +ve and situated close to the neurohypophysis. Little certain information is available about functional changes in these cells, although they have been examined in lamprey larvae made pale by constant illumination, but with no clear results. Various other changes have been described in the pars intermedia in relation to metamorphosis, migration, and spawning, but they are difficult to interpret.

The neurohypophysis of myxinoids is peculiar, consisting of a dorsoventrally flattened hollow sac above the pars intermedia region, which communicates with the third ventricle only by a narrow aperture. Nearly all the neurosecretory fibers from the NPO terminate in the highly vascular dorsal wall of this sac, there being few nerves or blood vessels in the ventral wall. Many of the axons contain granules varying in size from 100 to 200 mμ, together with much smaller vesicles. In lampreys, the neurohypophysis is merely a slight thickening of the floor of the infundibulum above the pars intermedia. It is composed of fibers from the NPO, many of which terminate around neurohypophysial blood vessels or between the ependymal cells. There is no evidence of neurohypophysial penetration between the adenohypophysial cells.

Exposure of larval lampreys to continuous light decreased the amount of neurosecretory material in the NPO and proximal axons, while continuous darkness had the reverse effect. Neurosecretory material became scarce during metamorphosis of *L. planeri*, and AF +ve material completely disappeared during gonad maturation. Sterba and Bruckner (1967) made ultrastructural studies on the neurohypophysis of *L. planeri* after hypothalamic lesions; they found that the axons degenerated, liberating the elementary granules which were phagocytosed by ependymal cells and then released into the third ventricle. The authors obscurely interpreted these phenomena as evidence for a normal feedback route from neurohypophysis to cerebrospinal fluid, as suggested by Knowles and Vollrath for the eel.

The blood supply to the cyclostome pituitary, which differs in the two groups, has been reviewed by Gorbman (1965b). In myxinoids, the adeno- and neurohypophysis are independently vascularized by branches of the internal carotid. In addition, in *Myxine glutinosa* the dorsal region of the neurohypophysis receives blood by a portal vessel from what appears to be a median eminence in the floor of the hypothalamus just anterior to the pituitary. A similar portal system has been observed in the hagfish, *Polistotrema*, supplying the neurohypophysis from a neurohemal area just behind the optic chiasma. No portal vessels or blood from the neurohypophysis seems to supply the adenohypophysis, the two components being largely separated by connective tissue. Blood leaves the adenohypophysis by a posterior vein which then passes through the neurohypophysis before leaving the gland. The neurohypophysis is also drained by a more anterior hypophysial vein.

In the lampreys, the pars distalis has an arterial supply and venous drainage separate from the neurohypophysis and pars intermedia. Several small capillaries from the internal carotid enter the pars distalis, and several small venules drain this region. The curious portal supply to the neurohypophysis seen in the hagfishes is absent from lampreys, this region being solely vascularized by a branch of the internal carotid which forms a vascular plexus between the neurohypophysis and the pars intermedia. Blood is drained from this region by a ventral hypophysial vein.

Obviously, these ancient animals have highly peculiar vascular supplies to the pituitary. In hagfishes there seems to be neither neural nor vascular links between the neurohypophysis and adenohypophysis, while in lampreys the pars distalis is similarly isolated from the neural component but there appears to be the possibility of neurohypophysial-pars intermedia communication via the nerve terminations on the vascular plexus between the two regions. The functional significance of the extraordinary hypothalamo-neurohypophysial portal link in myxinoids is totally obscure.

3

POSTERIOR LOBE OF PITUITARY

Studies of the neurohypophysis provide a good example of the interaction between mammalian and comparative physiology. They show clearly how each division can be of benefit to the other. It was Scharrer's pioneer work on the teleost neurohypophysis that introduced the concept of hypothalamic neurosecretion into mammalian research, and replaced the original idea of the posterior pituitary as a simple endocrine gland, by the concept of a complex neurohypophysial system, in which hormones, secreted by nerve cells in the hypothalamus, passed down the nerve axons, to be stored in the pars nervosa of the pituitary. In turn, studies of the mammalian neurohypophysial principles opened the way to new ideas concerning the nature of the hormones present in fish. Investigation of the pressor, oxytocic, milk-ejection, avian depressor, and antidiuretic activities of the mammalian pars nervosa culminated in the purification, analysis, and synthesis of oxytocin and arginine vasopressin. However, the demonstration of an analog, lysine vasopressin, in the pig, opened up the possibility of species variability in neurohypophysial peptides. It gave new significance to Heller's early demonstration that lower vertebrate pituitaries, including those of teleost fish, contained a neurohypophysial principle which was particularly potent in promoting water reabsorption in the frog. Soon, this "water balance factor" was shown to be arginine vasotocin, a molecule containing moieties from both oxytocin and arginine vasopressin, and it became clear that the lower vertebrates, including the fish, contained a fascinating family tree of interrelated principles. Perhaps the least satisfactory phase of neurohypophysial studies has been in the elucidation of the function of neurohypophysial principles in the metabolism of fish. Although concepts derived from mammalian studies have been useful, they may have clouded our view to some extent, since it is probable that the peptides have new and, as yet, undiscovered functions in aquatic vertebrates. Perhaps further work will allow the fish to introduce yet new ideas into mammalian physiology, this time new ideas concerning function. In this vein, it is particularly interesting that recent work has suggested that arginine vasotocin may be present in the mammal during its term of "aquatic" existence, i.e., during fetal development in the uterus.

Cyclostomes

Structure of the Neurohypophysis of the Cyclostomes

The hagfish (Myxiniformes) and the lampreys (Petromyzoniformes) hold a place of special importance in comparative studies, since they are the only living representatives of the first vertebrate class to be found in the fossil record. Already, they possess a relatively well developed neurohypophysial system, with only general tendencies which might appear to be primitive. This implies that comparative studies can examine only the final molding of the system, and its origins will remain obscure. However, the fact that it exists in both the two living groups of cyclostomes, which may have originated separately from ancestral Ostracoderms, suggests that it was present also in these common ancestors. If this diphyletic origin is correct, the hagfish and the lampreys may not be as closely related as is often assumed. This consideration, together with their long history of separate development, their clear specializations, and their degenerate features, makes it difficult to be certain which characteristics of their neurohypophyses are truly primitive, and which are developments peculiar to each group. Also, it suggests that they should be given separate consideration.

Myxiniformes

Despite the difficulties pointed out above, a number of workers have concluded that the Myxiniformes possess the most primitive pituitary of any living vertebrate. However, in his early studies, de Beer (1926) had suggested that their pituitaries might be degenerate; despite this, he was impressed by the development of the infundibular process and felt that it was better formed than in the lampreys. Unlike the lamprey, the infundibulum of the hagfish consists of a hollow, flattened sac, 1 mm long, and attached to the brain by a narrow stalk. Early accounts suggested that this process was poorly differentiated, but recent work has shown that all the components of a neurohypophysial system are present in both *Myxine glutinosa* and in *Polistotrema stouti*. The following general account of the hagfish neurohypophysis is based on the work of these recent authors.

In the hagfish, the paired preoptic nuclei consist of ill-defined clusters of cell bodies, located dorsally to a poorly developed optic tract. The more ventral cells ("parvocellularis") lie close to the preoptic recess and do not appear to contain neurosecretion. In contrast, the dorsal cells ("magnocellularis") contain small quantities of fine, perinuclear neurosecretory granules which stain with Astra blue, but which are not detectable by the classic chrome-hematoxylin-phloxin stain of Gomori. Unlike neurosecretory cells in many other lower vertebrates, the preoptic cells of the hagfish do not send dendrites to the cerebral ventricle; however, it should be pointed out that the ventricles are greatly reduced in these species. Neurosecretion may leave the nucleus by two possible pathways. The first is probably unique to the Myxiniformes: It is possible that neurosecretory material may pass directly from the nucleus into a portal system which leads to the neural lobe of the pituitary. This is suggested by the presence of accumulations of secretion between and within the capillary walls of a vascular plexus which lies directly beneath the preoptic nucleus; this plexus has been seen to drain to the neural lobe and pars nervosa. The second pathway is well established in many vertebrates;

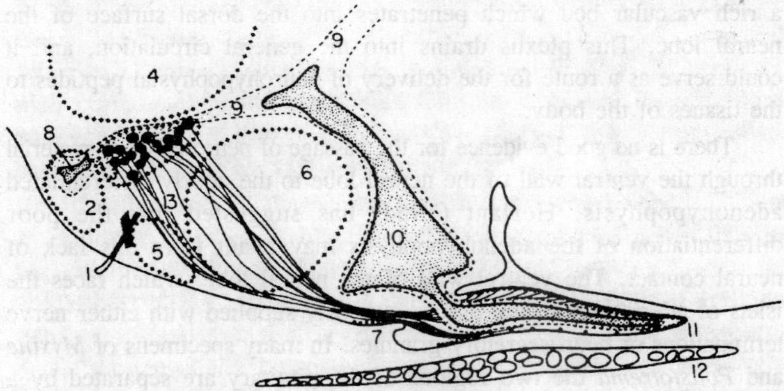

Fig. 3.1. The hypothalamus and pituitary of the hagfish, Myxine glutinosa, sagittal section. (1) Optic tract; (2) preoptic nucleus, pars parvocellularis or anterior region; (3) preoptic nucleus, pars magnocellularis; (4) primordium hippocampi; (5) postoptic nucleus; (6) optic commissure; (7) possible median eminence; (8) preoptic recess; (9) area of remnants of buried ependymal cells; (10) third ventricle; (11) infundibular process; (12) adenohypophysis; and (13) preoptico-hypophysial tract.

neurosecretion may leave the nucleus along the extremely thin processes which make up the preoptico-hypophysial tract of the hagfish. The axons disseminate widely at first. They are mostly empty of neurosecretion in this part of the tract, but fine droplets occur on rare occasions. The axons converge on the floor of the hypothalamus, and there is an accumulation of large droplets of neurosecretion, termed "Herring bodies," near to the origin of the infundibular process. This region is highly vascular, and both Olsson (1959) and Adam (1963a) have suggested that it represents the median eminence of the higher vertebrates. However, Gorbman et al. (1963) regard it as no more than the most anterior part of the neural lobe or pars nervosa. Some axons appear to form nerve endings in the narrow stalk of the infundibular process. However, most of the fibers of the preoptico-hypophysial tract continue to the end of the process, where they form an ill-defined pars nervosa within the terminal neural lobe. Olsson (1959) has likened this lobe to the equivalent structure in the embryo of the higher vertebrates. It is possible that some nerve fibers penetrate the inner ependymal lining of the neural lobe and deliver neurosecretory products into the infundibular cavity. However, the great majority of the neurosecretory fibers terminate in the dorsal wall of the lobe, mostly in the outer two-thirds of its thickness, and this might be regarded as the pars nervosa. Here, there is a relatively dense accumulation of neurosecretion. The neurosecretory droplets are closely associated with a rich vascular bed which penetrates into the dorsal surface of the neural lobe. This plexus drains into the general circulation, and it could serve as a route for the delivery of neurohypophysial peptides to the tissues of the body.

There is no good evidence for the passage of neurosecretory material through the ventral wall of the neural lobe to the poorly differentiated adenohypophysis. Herlant (1954) has suggested that the poor differentiation of the adenohypophysis may result from this lack of neural contact. The ventral wall of the neural lobe, which faces the islets of adenohypophysial tissue, is poorly supplied with either nerve terminations or neurosecretory granules. In many specimens of *Myxine* and *Polistotrema* the two regions of the pituitary are separated by a thick layer of connective tissue. This is penetrated only by a few, small blood vessels. However, Matty (1960a,b) has described specimens of *Myxine* in which there is intimate contact between the caudal region of the neural lobe and the adenohypophysis. This apparent contradiction of a number of other workers could result from individual variation, but it could also be explained by Adam's observation (1963a) that the connective tissue septum becomes thinned in specimens over 30 cm in length. If this degree of contact increased with age and length, it is possible that the neurohypophysis could come to influence the adenohypophysis in later life. However, Matty could find no evidence for the passage of neurosecretory droplets into the adenohypophysial tissues even though contact existed. Clearly, it is not possible to discount interactions between the two regions of the pituitary, but the present consensus is against their existence. It is worth noting that Gorbman et al. (1963) have pointed out that the neurohypophysial circulation of the hagfish is well adapted to the delivery of hormones to the general circulation.

Petromyzoniformes

The neurohypophysial system of the lamprey is markedly different from that of the hagfish. Unlike the hagfish, the lampreys do not develop a clear infundibular process. Nevertheless, studies of *Petromyzon marinus*, *Lampetra lamottei*, and *L. planeri* have shown that a complete neurohypophysial system is present in the slightly thickened floor of the hypothalamus. The preoptic nucleus is well marked, although less prominent and less vascular than in the teleost fish. It lies in a more rostral position than that of the teleosts, and it is found lining the preoptic recess, anterior to the optic chiasma. The preoptic cells are bipolar and of relatively uniform size; there is no division into parvocellular and magnocellular regions (*Petromyzon marinus*). Even in the youngest ammocoete larva so far examined, these cells contain neurosecretion, but it is different from that of the hagfish, since it will stain by both Gomori's chrome-hematoxylin-phloxin stain, and with aldehyde fuchsin. The preoptic cells of larval and adult *Petromyzon marinus* may lie close to the ventricle itself, or they may send short dendrites containing neurosecretion between the ependymal cells, to the cerebrospinal fluid. This is not found in the hagfish. In addition to these connections with the ventricle above, a few preoptic nerve fibers appear to make direct or indirect connections with the adenohypophysis below. Roth (1956) has suggested that in adult *Petromyzon marinus* some nerve fibers which enter the region close to the preoptic recess may contact capillaries which supply the adenohypophysis. Oztan and Gorbman (1960b) have reported that in the ammocoete larva of the same species, a small number of preoptic fibers terminate in the vascular connective tissue between the rostral adenohypophysis and the brain. It is even possible that these fibers,

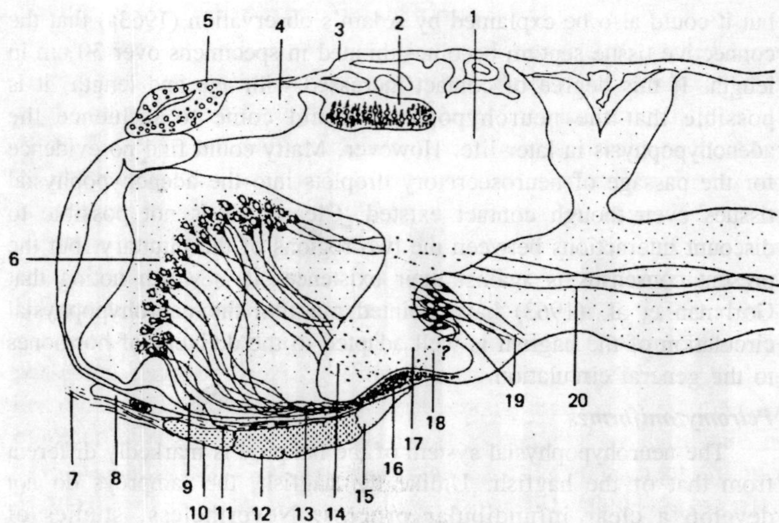

Fig. 3.2. The brain, hypothalamus, and pituitary of the ammocoete larva of the lamprey, Petromyzon marinus, sagittal section. (1) Third ventricle; (2) subcommissural organ; (3) choroid plexus; (4) habenula; (5) pineal body; (6) preoptic nucleus; (7) nasopharyngeal stalk; (8) blood vessel; (9) optic chiasma; (10) adeno-hypophysis; (11) preoptic neurosecretory axons ending close to adenohypophysis; (12) adenohypophysis (proximal zone); (13) preoptico-hypophysial tract (ventral division); (14) preoptico-hypophysial tract (lateral division); (15) pars intermedia; (16) pars nervosa; (17) infundibular cavity; (18) axons from posterior hypothalamic nucleus; (19) posterior hypothalamic nucleus; and (2) preoptic axons connecting to the midbrain.

which bear beaded neurosecretory granules, may penetrate the adenohypophysis itself. In addition, a few preoptic fibers appear to connect with the hindbrain. However, as Roth (1956) has pointed out in *Petromyzon marinus*, the majority of the axons of the preoptic cells are directed caudally, toward the pars nervosa. These axons carry beaded droplets of neurosecretion and form into a discrete preoptico-hypophysial tract over the rostral region of the adenohypophysis. At this point, the tract is relatively avascular, and it is separated from the adenohypophysis by an almost continuous sheet of connective tissue. The tract runs within the ventral hypothalamic wall. This wall is relatively thick and full of neurosecretion in *Lampetra planeri*, but in *Petromyzon marinus* it is thin, and contains few droplets or nerve endings. The axons of the preoptico-hypophysial tract terminate caudally in a slightly thickened area of the hypothalamic wall. They enter in a rostro-caudal direction, and also laterally, and they may be joined by a few axons from neurosecretory cells located in a posterior hypothalamic nucleus (nucleus lateralis tuberis?). In *Lampetra planeri* the thickened area is broadly continuous with the floor of the brain, but in *Petromyzon marinus* it has a distinctly globular structure with a central septum at its caudal extremity. Although it is not as clearly demarkated as in a mammal, this region may be regarded as the neural lobe or pars nervosa since it appears to store neurosecretion. A few of the neurosecretory axons which enter this region carry beadlike droplets of neurosecretion between the ependymal cells which line the ventricle. These ependymal cells, which contain aldehyde-fuschin-positive granules, send down processes into the underlying neural tissue and form a structural lattice for the pars nervosa. Occasionally, these ependymal cells appear to sink into the pars nervosa, where they have been described as pituicytes. Although a few neurosecretory fibers associate with the ependyma, it is clear that most terminate in swollen accumulations of neurosecretion, like Herring bodies, and these are grouped around blood vessels. Although the ventral surface of the pars nervosa interdigitates with the pars intermedia to give a serrated appearance, there is no evidence that any neurosecretory axons penetrate the intermedia tissue (*Petromyzon marinus*). The border between the two regions is marked by thin connective tissue, which is liberally supplied by a reticulum of capillaries (*Petromyzon marinus, Lampetra planeri*). This reticulum is filled and drained by blood vessels which are independent of those concerned with the rostral regions of the adenohypophysis. Since there is no observable portal system in the lamprey pituitary, and these vessels both supply the pars nervosa and penetrate the pars intermedia, they are a potential vascular link between the two regions of the pituitary. It may be significant that van der Kamer and Schreurs (1959) have noted that these vessels enlarge during the metamorphosis of the ammocoete larva of *Lampetra planeri*. However, these same blood vessels also empty into the general circulation of the head, so that they form a potential route by which neurohypophysial peptides could reach all parts of the body.

Nature of the Neurohypophysial Principles of the Cyclostomes

Myxiniformes

There is little knowledge of the principles present in the hagfish. This is because of the disappointingly low activities found in the glands, together with the difficulties of obtaining these deep-water animals. Such studies as exist are confined to one species, *Myxine glutinosa*. Herring's remarkable early work reported the presence of a vasopressor principle in this species. It was almost half a century before Adam observed a frog water-balance effect when extracts from *Myxine* were tested on *Bufo viridis*. In 1964, Follett and Heller were able to detect oxytocic and natriferic activity in an extract of 150 glands, but they failed to find a vasopressor effect (1964a). The total oxytocic activity obtained amounted to only 74 mU, and since no more than one quarter of this showed the lability to sodium thioglycollate treatment typical of neurohypophysial principles, they deduced that the hagfish pituitary contained no more than 0.13 mU/gland. Although it is possible that the reagent was partly expended on extraneous materials in so crude an extract, it is clear that the hagfish pituitary contains only traces of an oxytocic agent. The detection of frog water balance activity, together with oxytocic and natriferic effects, suggests the possibility that arginine vasotocin is present in the hagfish, as well as in the lampreys, but the present evidence is inadequate for anything more than speculation.

Petromyzoniformes

The neurohypophysial principles of the lampreys have received more attention than those of the hagfish. Herring (1913) showed that extracts from *Petromyzon fluviatilis* could cause a weak antidiuretic effect in the cat. These early studies were extended by Lanzing in 1954, when he detected frog water-balance and oxytocic activities in *Lampetra fluviatilis*. However, it was doubtful whether the agent which stimulated the guinea pig uterus was a true neurohypophysial peptide, since it was resistant to sodium hydroxide inactivation. In 1955, W. H. Sawyer was able to demonstrate that pituitaries from *Petromyzon marinus* contained oxytocic and vasopressor activities which were labile to sodium hydroxide treatment, and, in a series of studies carried out by W. H, Sawyer et al. (1959, 1960, 1961), milk-ejection, antidiuretic, frog bladder and hen oviduct stimulating actions were added to the properties of the extracts. Careful quantitative comparisons of these activities suggested that the extracts contained arginine vasotocin, a molecule which shared the ring structure of oxytocin with the terminal side chain of arginine vasopressin. This peptide might well serve as a potential "primitive" neurohypophysial principle. Purification of *Petromyzon marinus* extracts by gel filtration and the use of carboxymethyl cellulose columns yielded an active eluate with pharmacological properties similar to arginine vasotocin. A small "notch" in the eluted activity peak left some slight possibility that two similar peptides could coexist in the extracts. The purification procedure did not separate any oxytocinlike neutral peptide. However, it is possible that small quantities of a neutral peptide could have escaped detection during pharmacological studies, and these might have been lost during the prolonged storage of the extract prior to purification. In essentially similar pharmacological studies on *Lampetra fluviatilis*, Follett and Heller (1964a) showed that oxytocic, vasopressor, and natriferic activities were also in agreement with the presence of arginine vasotocin. Paper chromatography in butanol-acetic acid-water, 4:1:5, showed that the major activity peak ran with an R_f value of 0.25-0.35, which was similar to that of arginine vasotocin. Although, milk-ejection effects and antidiuretic activity were detected at R_f values between 0.5 and 0.8—the approximate position of oxytocinlike peptides—Follett and Heller did not consider that this was evidence for the presence of an oxytocinlike principle. However, their findings suggest that there should be some caution before the possibility of an oxytocinlike peptide is eliminated from cyclostome physiology. Most work has been done on adult animals during their spawning migration, and W. H. Sawyer (1961) has pointed out that oxytocinlike peptides might exist during other stages of the life cycle. The histological studies of van der Kamer and Schreurs (1959) have shown that neurosecretory material is remarkably dense in parts of the neurohypophysis during the ammocoete stage, but parallel pharmacological studies are lacking.

There have been no chemical studies of the structure of any cyclostome principal. Nevertheless, all present evidence indicates that the cyclostome pituitary contains an overwhelming preponderance of arginine vasotocin and that some cyclostomes may well be unique in possessing a single neurohypophysial principle in their pituitaries.

Actions of Neurohypophysial Principles in Cyclostomes

Myxiniformes

Studies of the actions of neurohypophysial principles in the hagfish are few and brief. It is possible that the peptides influence salt and water metabolism. Chester Jones et al. (1962), working with *Myxine glutinosa*, determined the effects of neurohypophysial principles on hagfish whose body fluids had been diluted or concentrated by immersion in diluted or concentrated seawater. Hagfish pituitary extracts, equivalent to 10 glands, or possibly 1.3 mU of oxytocic activity, were injected daily for 4 days into hagfish adapted to diluted (70%) seawater. The treatment resulted in a rise in the serum sodium level. A mammalian oxytocin-vasopressin preparation (Pituitrin) was injected at high dose levels (1000 mU), daily for 5 days, into hagfish adapted to concentrated (165%) seawater. This treatment resulted in a fall in serum sodium level. It is dangerous to generalize from such scanty data, but it seems possible that neurohypophysial peptides tended to return body sodium levels toward the natural value for normal seawater, presumably by affecting sodium fluxes. Although the hagfish is incapable of maintaining its internal environment in the face of external changes in osmolarity, this effect on sodium levels might delay, and therefore buffer, the changes which follow slight, transitory variations in the salinity of its relatively constant, deep-water environment. However, the original authors draw the careful conclusion that the neurohypophysial peptides can affect the relationship between the extracellular and intracellular distribution of water and electrolytes. There is a hint that the hagfish's own principle(s) are more potent in the hagfish than those of the mammals. A second possible effect of neurohypophysial principles on the metabolism of the hagfish is an influence on water balance. Adam (1963a) observed that in *Myxine glutinosa*, the injection of high doses of a mammalian vasopressin preparation (Pituifral, 100-1000 mU) resulted in up to 7% rise in total body water. Clearly, there is need for more extensive investigations, with the use of the principle(s) native to the hagfish pituitary.

Since water, salts, and gases might be exchanged readily by the gills, it is interesting to note that the experiments of Somlyo and Somlyo (1968) have suggested that neurohypophysial peptides might influence the branchial circulation of the blood. They showed that isolated strips taken from the ventral aorta of *Eptatretus stautii* contracted to extremely low doses of synthetic arginine vasotocin (10^{-10} *M*), and that higher doses of oxytocin (0.1-100.0 mU/ml) and arginine vasopressin (1.0-100.0 mU/ml) would produce a similar response. The isolated dorsal aorta behaved in a similar way, but it was less sensitive, and it relaxed more readily. There are no studies of more general effects of neurohypophysial principles on the blood pressure or circulation of the hagfish.

At present, it is not possible to reach dependable conclusions on the function of the neurohypophysis of the hagfish. Further study is made more urgent by the demonstration of Gorbman et al. (1963) that the hagfish neurohypophysis is unlikely to be concerned in the local control of the adenohypophysis and is specially adapted to pour its secretions into the systemic circulation.

Petromyzoniformes

Although experiments involving hypophysectomy have suggested that the pituitary of the freshwater lamprey was not essential to life, the extent to which the pars nervosa had been removed is uncertain, and, in any case, the intact preoptic nuclei might have continued to supply the necessary hormones (*Lampetra planeri*, *L. fluviatilis*).

Direct studies with neurohypophysial peptides have been confined almost entirely to the freshwater lamprey, *Lampetra fluviatilis*, and they have produced a number of negative results. The injection of pure arginine vasotocin in doses of 36 mμ moles/kg failed to cause any water-balance effect, and injections of 2.2 μg failed to change the volume of urine production over a 6-hr period. However, it is possible that the lamprey neurohypophysis may influence sodium metabolism, and this is reminiscent of the possible effects in the hagfish. The injection of a number of neurohypophysial peptides, including arginine vasotocin, into the peritoneum of *Lampetra fluviatilis*, resulted in an increase in the rate of sodium loss from the fish into its external medium. The site of action was in part the kidney, since injection of arginine vasotocin increased the sodium and potassium concentrations of the urine, and it is possible that a depression of tubular sodium reabsorption may be one factor involved. Although all the peptides tested provoked sodium loss, a comparison of equimolar doses suggested that arginine vasotocin, the natural principle of the lamprey, was only one-fifth as potent as mammalian oxytocin, on a weight basis. If comparisons were made on the basis of the oxytocic activity which was injected, arginine vasotocin was slightly less effective than oxytocin and considerably weaker than the potent 4 Ser, 8 Ile oxytocin—which is the natural oxytocinlike peptide of the teleost fish. This unexpected weakness of arginine vasotocin must lead to speculation on the possible existence of a neutral, oxytocinlike peptide in the lamprey pituitary. This second peptide might be present only during the marine period of existence, for Bentley and Follett (1962) have suggested that the actions of the principles on sodium metabolism would be of greatest use during the marine phase of the life cycle. Further, Dodd et al. (1966) have pointed out that the high doses required to produce a sodium response could reflect a reduced tissue sensitivity which had occurred when the lamprey first entered freshwater. Morris (1960) has suggested that there is a breakdown of osmotic controls at this time. This could also be connected with the low levels of neurohypophysial peptides found in pituitaries taken from lampreys adapted to freshwater (1.2 mU/gland, *Lampetra fluviatilis*). In fact, the doses of neurohypophysial principles needed to evoke a sodium loss are greater than the total hormonal activity available in the lamprey's pituitary. However, it must 130 remembered that a low hormone content in the pituitary does not necessarily imply a low rate of loss into the circulation, since the stored peptide is only a reflection of the balance between hypothalamic supply and pituitary loss. This same consideration may account for the failure of experiments to demonstrate a significant difference between glands dissected from lampreys living in freshwater and those immersed for a short period in more concentrated saline.

A completely different action of neurohypophysial peptides has been suggested by the fact that arginine vasotocin, in doses similar to those which affect sodium metabolism, will cause a rapid rise in blood sugar and in muscle glycogen. An injection of arginine vasotocin amounting to 1.2 mμ moles, or 150 mU oxytocic activity, resulted in at least 40% rise in the blood glucose level, possibly by a fat mobilizing action. The dose used was high, and the same considerations which were dealt with under sodium metabolism apply here. Again, the physiological significance of the response is uncertain.

The lamprey neurohypophysis may be of local as well as of systemic importance, but the only evidence available, at present, is from histology. W. H. Sawyer et al. (1960) have pointed out that the close relationship between the neurohypophysis, the intervening blood capillaries, and the pars intermedia suggest that neurohypophysial peptides could pass to the adenohypophysis and modulate its function. It is possible that this action could be of special importance at metamorphosis and spawning. Van der Kamer and Schreurs (1959) have shown the presence of a considerable accumulation of neurosecretion in the ammocoete larva of *Lampetra planeri*; however, there is a dramatic loss of this material at metamorphosis. During metamorphosis, neurosecretion is found mainly in the nerve terminals which border capillaries that enter the adenohypophysis. It is also found in the fibers which reach the ependyma. Metamorphosis is accompanied by activity of the pars intermedia, and the lamprey takes on an adult colour pattern. After spawning, the neurohypophysis is almost empty of granules. These observations are highly suggestive of involvement of the neurohypophysis in metamorphosis and spawning, but they do not prove that there is a direct relationship. There is a small possibility that the changes in neurosecretion are linked to the emergence of the larva from the dark mud. Oztan and Gorbman (1960a,b) have seen that the distribution of neurosecretion in *Lampetra lamottei* and *Petromyzon marinus* is influenced by light. When these species are placed in continuous darkness there is a dense accumulation of neurosecretion in the preoptic nucleus and its axons. When the lampreys are returned to continuous light, the material is depleted. However, unlike the changes seen in metamorphosis, the nerve terminals of the pars nervosa are not affected. It is possible that these various observations indicate a neurohypophysial control of the pars intermedia, but more direct information is needed before any conclusions can be drawn.

CARTILAGINOUS FISH: ELASMOBRANCHS

Structure of the Neurohypophysis of the Elasmobranchs

The pituitary of the elasmobranchs appears to be markedly different from that of the cyclostomes, and it is undoubtedly more advanced. However, there are a few surprising aspects which are reminiscent of both the hagfishes and the lampreys.

As early as 1685, Collins had illustrated the pituitary of the "skait"; he wrote, "The lower region of the brain of a skait seemeth to be composed of three ranks of processes, and an odd one." The "odd one" was an unusually large, spherical, centrally located process which lay beneath the caudal extremity of the saccus vasculosus; it was clearly the neurointermediate lobe of the pituitary. Despite this early start, evidence for the existence of a neurohypophysis in the elasmobranchs was slow to take shape. Many early workers failed to find any area which could be termed a pars nervosa. A few investigators accepted the thin infundibular lamina of the ventral surface of the brain as the only possible candidate for the neural lobe. However, in 1926 both Pokorny and de Beer observed areas of neuroglia-type cells within the pars intermedia of *Raia clavata* and other species, and this was confirmed by Howes in 1936. The reason for the early difficulties became clear when Scharrer (1952) used Gomori's chrome-hematoxylin-phloxin technique to demonstrate the presence of a complete neurohypophysial system in *Scyliorhinus stellaris* (syn. *Scyllium stellare*). He showed that the pars nervosa was a diffuse structure which penetrated throughout the tissue of the pars intermedia. Soon afterward, these observations were extended to *Raia clavata*, *Dasyatis marinus*, *Torpedo ocellata*, and *Scyliorhinus caniculus*. At about the same time, van der Kamer and Verhagen (1954, 1955) gave a detailed description of the histology of the pars nervosa of *Scyliorhinus caniculus*. They observed the presence of both pituicytes, and of secretory "parenchymatous pituicytes," which they considered to be possible sources of hormones. In 1960, Perks, Dodd, and Dodd, working on the same species, confirmed the general structure of the neurohypophysial system by the use of the chrome-hematoxylin-phloxin stain, and showed that the neurosecretory droplets were probably rich in sulfur, since they stained by the performic acid-Alcian blue method of Adams and Sloper (1956). The vacuolated parenchymatous pituicytes did not show any evidence for a high sulfur content, and therefore they were unlikely to be a source of neurohypophysial peptides. In 1962, Braak was able to trace the neurosecretory tracts of *Spinax niger* by the elegant technique of staining and clearing whole mounts. In the following year, Knowles (1963) applied electron microscopy to the study of the diffuse neural lobe of *Scyliorhinus stellaris*. However, the most important advance of recent years has been the demonstration of a hypothalamo-hypophysial portal system in the elasmobranch pituitary. Meurling (1962, 1963, 1967a) has made careful comparisons of both the portal system and the neurohypophysis in different species, and the following general description is based largely on his excellent work.

The preoptic nucleus of the elasmobranchs is found dorsal to the optic chiasma and close to the ependymal lining of the ventricle. It is shorter than that of most teleosts, and its position varies from being relatively rostral in the sharks (especially in *Squalus acanthias*), to being remarkedly caudal in the skates. The cells are often found lying with their long axes parallel to the ventricular lining (*Scyliorhinus caniculus*). They are relatively uniform in size, and there is no division into a pars magnocellularis and a pars parvocellularis, as seen in the teleosts. The preoptic cells are usually large and distinct, as in *Etmopterus spinax* and *Scyliorhinus caniculus*, but in some species such as *Squalus acanthias* they are less well marked. They contain innumerable small granules, which stain by Gomori's chrome-hematoxylin-phloxin method, and by other so-called neurosecretory stains (e.g., *Scyliorhinus stellaris*). The droplets appear to contain a high sulfur content (*Scyliorhinus caniculus*). In *Scyliorhinus caniculus* there is a great variation in the amount of neurosecretion present in the preoptic nucleus, and indeed, throughout the entire neurohypophysis. If the neurointermediate lobe of the pituitary is removed, or the preoptico-hypophysial tract is cut, the preoptic nucleus becomes completely empty of neurosecretory granules (*Scyliorhinus caniculus*). In normal specimens different cells within the same nucleus appear to be in different stages of secretion (*Scyliorhinus caniculus*, *Scyliorhinus stellaris*). Studies with the light microscope have suggested that neurosecretory granules appear in the walls of peripheral cytoplasmic vacuoles, and then spread throughout the cytoplasm, becoming particularly dense in the poles of the cells (*Scyliorhinus caniculus*, *S. stellaris*). Beads of neurosecretion can be found in dendrites which pass to the third ventricle, but most appear to leave the cells along the thick cellular processes, which

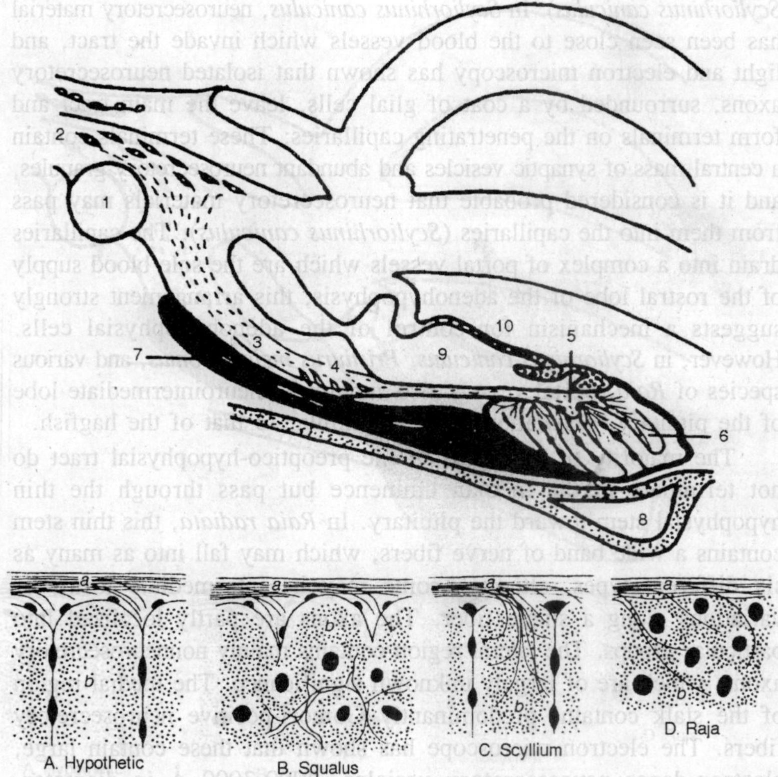

Fig. 3.3. The hypothalamus and pituitary of the elasmobranchs; diagrammatic representation. (1) Optic chiasma; (2) preoptic nucleus; (3) preoptico-hypophysial tract; (4) median eminence, (5) pars nervosa; (6) pars intermedia; (7) adenohypophysis; (8) ventral lobe of the pituitary; (9) infundibular cavity; and (10) saccus vasculosus.

narrow down to form the unmyelinated fibers of the preoptico-hypophysial tract. The axons of this tract run down behind the optic chiasma in a diffuse manner; at first they lack neurosecretory granules and give only a uniform staining, which makes them difficult to follow. However, in the caudal postoptic lamina the axons converge and form a discrete tract close to the midline (*Squalus acanthias*). Here the tract shows densely packed neurosecretory granules and Herring bodies in many—but not all—of its fibers (*Scyliorhinus stellaris*, *Dasyatis marinus*, *Scyliorhinus caniculus*). In this region, portal capillaries penetrate the tract (*Scyliorhinus caniculus*, *S. stellaris*, *Raia asterias*, *R. punctata*, *R. undulata*, *Scyliorhinus caniculus*), and a median eminence is formed (*Scyliorhinus caniculus*, *Squalus acanthias*, *Raia batis*, *R. radiata*,

Scyliorhinus caniculus). In *Scyliorhinus caniculus*, neurosecretory material has been seen close to the blood vessels which invade the tract, and light and electron microscopy has shown that isolated neurosecretory axons, surrounded by a coat of glial cells, leave the main tract and form terminals on the penetrating capillaries: These terminals contain a central mass of synaptic vesicles and abundant neurosecretory granules, and it is considered probable that neurosecretory materials may pass from them into the capillaries (*Scyliorhinus caniculus*). The capillaries drain into a complex of portal vessels which are the sole blood supply of the rostral lobe of the adenohypophysis; this arrangement strongly suggests a mechanism for control of the adenohypophysial cells. However, in *Scyliorhinus caniculus*, *Pristiurus melanostomus*, and various species of *Raia*, portal vessels also supply the neurointermediate lobe of the pituitary, a situation strangely similar to that of the hagfish.

The majority of the axons of the preoptico-hypophysial tract do not terminate in the median eminence but pass through the thin hypophysial stem toward the pituitary. In *Raia radiata*, this thin stem contains a wide band of nerve fibers, which may fall into as many as six different types, although some may be intermediate forms or variations along a single fiber. The axons are partly localized into particular groups. The lateral region contains mainly nonneurosecretory axons, which are of largely unknown significance. The medial region of the stalk contains predominantly Gomori-positive neurosecretory fibers. The electron microscope has shown that these contain large, electron-dense neurosecretory vesicles, 2000-3000 Å in diameter, probably comparable to the Type A elementary vesicles described in the neurointermediate lobe of *Scyliorhinus caniculus*. In addition, there are axons which contain smaller, electron-dense vesicles, 1000-1300 Å in diameter, which have a distinct external membrane, and may be comparable to the Type B elementary vesicles of *Scyliorhinus caniculus*. In some cases, these latter fibers form terminal swellings within the stalk. It has been suggested that they may affect the nearby pars medialis of the adenohypophysis, or perhaps liberate neurovascular transmitters into the nearby portal vessels, which could mediate an influence on the pars intermedia. As the neurosecretory fibers continue in a caudal direction, glial elements and secretory cells become less frequent, and it is probable that the secretory cells are lost before the axons of the preoptico-hypophysial tract enter the neurointermediate lobe of the pituitary. Here, axons bearing granules and Herring bodies, and some devoid of secretion (*Dasyatis marinus*) pass below the single or multiple layers of ependymal cells which line the ventricle (*Squalus acanthias*).

The overlying ependymal cells send down fibers which divide the tract into bundles (*Squalus acanthias*). In *Squalus acanthias* and *Scyliorhinus caniculus* some ependymal cells which contain vacuoles are found enclosed within the tract itself. In some species the neurosecretory axons terminate in a fairly well defined pars nervosa, but in others the terminations penetrate diffusely throughout the pars intermedia of the pituitary. The well-defined pars nervosa of *Squalus acanthias* may be a primitive feature. In this species the neurosecretory axons carry variable numbers of granules of different sizes, and they are sometimes in the form of beads. The axons are confined mainly within a lamina of variable thickness which covers the dorsal surface of the pars intermedia. This lamina sends down solid and tubelike processes into the intermedia tissue, and these projections, together with the dorsal neural covering, constitute the major portion of the pars nervosa. In the primitive sharks, *Hexanchus*, *Heptanchus*, and *Chlamydoselachus*, similar rodlike and tubelike processes are well developed. Within the pars nervosa of *Squalus acanthias* most axons terminate in an outer layer of ependymal fibers which follows the boundary with the pars intermedia. This layer contains a dense accumulation of neurosecretion. It contacts an external reticular membrane, which encompasses a network of capillaries. This neural lobe plexus forms the boundary between the pars nervosa and the pars intermedia. It drains into the sinuses of the intermedia tissue and could form a neurovascular link between the two areas of the pituitary. This is reminiscent of the situation in the lampreys. However, neurosecretory axons do enter the pars intermedia of *Squalus acanthias* to a small extent. Sometimes isolated axons penetrate the boundary membrane and take an irregular course between the intermedia cells. More often, and particularly in the neural processes, groups of neurosecretory axons pass through gaps in the reticular membrane and run between the intermedia cells or within the connective tissue which divides the cells into cords. Other species show a greater penetration by the pars nervosa. *Etmopterus spinax* is essentially similar to *Squalus acanthias*, except that the pars nervosa is covered by a connective tissue membrane which contains extensive fenestrations in the posterior region of the gland. Here, nerve axons penetrate the pars intermedia; some are neurosecretory and form occasional terminal swellings against the cells, but others carry no neurosecretion and terminate in a nerve plexus around the blood sinusoids of the pars intermedia. In *Scyliorhinus caniculus*, *S. stellaris*, and *Mustelus mustelus*, the neural lobe is still recognizable, but there is no membrane and no typical, continuous neural lobe blood plexus to limit its boundary. In *Scyliorhinus caniculus* nerve fibers penetrate between the intermedia cells, and neurosecretion of high sulfur content is found packing closely around these cells. Electron microscopy has shown that the neurosecretory nerves make close contacts, suggestive of synapses, with the intermedia cells. In *Scyliorhinus stellaris* (syn. *Scyllium stellare*) a high proportion of the neurosecretory axons enter the pars intermedia and secretion is stored everywhere between the cells. Electron microscopy has shown that the neurosecretory fibers can be divided into two types. The first, Type A, contains large electron dense vesicles and makes contact with the "synthetic" region of the intermedia cells. The second. Type B, contains small, irregular neurosecretory elements and terminates on the "storage and release" region of the cells. This suggests the possibility of a separate control of synthesis and release from the intermedia cells. In *Pristiurus melanostomas* there is an intimate fusion of neural and intermediate components, and in many places a neural lobe is no longer distinguishable; no neural lobe plexus has been described. In the skates and rays (*Raia batis*, *R. oxyrhynchus*, *R. fullonica*, *R. radiata*, *R. clavata*, *Dasyatis* sp., and *Torpedo* sp.) there is complete fusion of neural and intermediate elements. The only region which could possibly be termed a neural lobe is the small anterior area of the infundibular floor, where the preoptico-hypophysial tract enters the neurointermediate lobe. There is no neural lobe plexus, but blood vessels are found close to the ventricular surface, and the entire neurointermediate lobe contains a rich plexus of sinuses. A combination of light and electron microscopy has suggested that the bundles of nerve fibers which penetrate between the strands of intermedia cells in *Raia clavata* and *Trygon pastinaca* are made up of both neurosecretory axons and nerve fibers which do not carry such granules. The nerve axons are free of connective tissue or blood vessels, but they are accompanied by neuroglia-type pituicytes. These pituicytes are closely associated with the nerve terminals. The neurosecretory axons terminate as large swellings, which contain irregular mitochondria, vesicles resembling synaptic vesicles, and fine "elementary vesicles" which represent the neurosecretory granules of light microscopy. These elementary vesicles are 1100-2800 Å in diameter, and it seems probable that they possess an outer lipoprotein membrane with electron dense material inside. Within the nerve terminals there are similar structures with contents of lower electron density, and others which appear empty, except for a few granules; it is probable that these represent various stages in the loss of neurosecretory materials. Since no granules can be found outside the terminals, Polenov and Belenki (1965) suggest that the neurosecretory neurons lose components of low molecular weight from their elementary vesicles. Many of the neurosecretory terminals make close contact with intermedia cells; none appears to associate directly with the blood sinusoids in *Raia clavata* and *Trygon pastinaca*.

There is no doubt that neurosecretory nerves make direct contact with the cells of the pars intermedia in many species of elasmobranchs. However, vascular connections could also be important. The neurosecretory axons deposit dense neurosecretion close to the neural lobe blood plexus of species such as *Squalus acanthias* and *Etmopterus spinax*. The neural lobe plexus drains into the sinusoids of the pars intermedia, and this could allow neurosecretory materials to pass from the pars nervosa into the pars intermedia. In those species which lack a neural lobe plexus, direct neural contacts may be more important, but it is interesting to note that these are also the species which possess a neurointermediate lobe component of their pituitary portal system, so that humoral materials could be supplied directly from the median eminance. In *Scyliorhinus caniculus*, the neurointermediate lobe receives its entire blood supply from veins which leave the brain, and there is no direct arterial supply. Neurosecretory materials may not only reach the sinuses of the pars intermedia, but they may pass on from the sinuses, through the interorbital vein, into the general circulation. However, in some species, such as *Raia clavata*, loss into the systemic circulation does not appear to be as easy, since there is no neural lobe plexus, and the neurosecretory axons do not appear to contact the blood sinusoids of the pars intermedia. Nevertheless, the nerve terminals may well be sufficiently close to the rich blood sinuses of the intermedia tissue for their active agents to reach the general circulation.

Nature of the Neurohypophysial Principles of the Elasmobranchs

Early work on the biological activities of the elasmobranch neurohypophysis gave inconclusive results, mainly because of the low levels of activity which were present in the pituitary gland. Herring failed to find pressor effects or consistent changes of urine flow after the injection of skate pituitary extracts into cats. However, he obtained an oxytocic action on the isolated rat uterus after the addition of extracts from the pituitary of *Raia clavata*. His most remarkable observation was that of a rapid ejection of milk, after the intrajugular injection of an extract of skate pituitary into a lactating cat. At the time, he did not recognize this milk-ejection effect as a property of a neurohypophysial principle. In 1925, Hogben and de Beer showed that skate pituitary extracts produced a weak oxytocic effect on the isolated guinea pig uterus. Although Hogben (1925) found that the extract caused a weak depressor effect when injected into a duck, he did not emphasize that this was evidence for an elasmobranch oxytocinlike principle. In 1936, Geiling and Le Messurier gave a brief report of "a suggestion" of pressor activity, and an "inconclusive" antidiuretic effect from extracts of the pituitaries of dogfish and skate (*Squalus acanthias*, *Raia sp*.?). However, in 1941, Heller (1941a) detected and assayed a low level of antidiuretic activity in glands from European dogfish and skate (*Scyliorhinus caniculus*, *Raia* sp.?). Nevertheless, by 1950, Waring and Landgrebe had failed to confirm the presence of either oxytocic or vasopressor activities in glands taken from living elasmobranchs, and they suggested that the earlier positive results were caused by the presence of histamine. However, this suggestion was not in full agreement with all the known facts. It was not supported by Herring's early observation of milk-ejection activity nor by his detection of oxytocic effects on the isolated rat uterus, which was known to be insensitive to histamine. It was made even less likely when Maetz et al. (1959) observed both oxytocic and natriferic activity in extracts from the pituitaries of *Scyliorhinus caniculus* and *Hexanchus griseus*.

The consensus of these different observations suggested to Perks, Dodd and Dodd (1960; Perks, 1959) that the elasmobranchs might contain an oxytocinlike principle. For this reason they examined elasmobranch neurointermediate lobe extracts by a combination of different biological assays, and determined the properties of the active agents by the use of inhibitors known to destroy mammalian neurohypophysial principles. They detected milk-ejection, oxytocic, and antidiuretic activities in extracts from *Squalus acanthias*, *Scyliorhinus caniculus*, *Raia clavata*, and *Raia batis*. The activities were relatively weak. The oxytocic activities ranged between 7.0 and 54.2 mU/mg acetone dried powder, and the highest value so far reported is only 58.5 ± 10.8 mU/mg, for *Dasyatis sabina*. Pressor effects could not be detected. The active agents were distributed preferentially in the neurointermediate lobe. They shared with oxytocin inactivation by sodium thioglycollate, sodium hydroxide, ultraviolet light, and chymotrypsin. In contrast to the mammalian vasopressins, the elasmobranch oxytocic agent was resistant to trypsin, and therefore it was unlikely to contain arginine or lysine. Despite the general parallel with the properties of oxytocin, there were quantitative differences in the ratios of different biological activities. In *Squalus acanthias*, *Scyliorhinus caniculus*, *Raia clavata*,

R. batis, *Dasyatis sabina*, *Sphyrna mokarran*, *Eulania milberti*, and *Carcharinus leucas* the ratios of milk-ejection: oxytocic: antidiuretic activities were relatively consistent, and averaged 3.4: 1:0.03. This contrasted with ratios of approximately unity for mammalian (ox) pituitary powders, and of 1:1:0.0067 for synthetic oxytocin. This discrepancy between the elasmobranch and mammalian extracts was confirmed in *Squalus acanthias* by W. H. Sawyer et al. (1961). These workers extended the activities which could be found in the extract to avian depressor activity, frog bladder activity, and stimulation of the hen oviduct. In addition, they detected a weak pressor effect; although no pressor activity had been found by Perks et al. (1960) in their earlier experiments, it is probable that this was masked by the strong vasodepressor substances which were present in their extracts. Sawyer and his co-workers found a twofold potentiation of the oxytocic effect by the presence of magnesium ions; again, this indicated the presence of a peptide other than oxytocin. It was clear that the elasmobranchs container a new oxytocinlike principle of high milk-ejection potency, free, or nearly free, of any vasopressor peptide: This was the first demonstration of a naturally occurring, neutral analog of oxytocin.

The pharmacological studies left the possibility that the observed activities could be the result of a mixture of oxytocinlike peptides. Heller and his co-workers attempted to resolve this possibility by paper chromatography. They subjected extracts from *Squalus acanthias*, *Scyliorhinus caniculus*, *Raia clavata*, and *Negaprion brevirostris* to paper chromatography in butanol:acetic acid:water, 4:1:5. The oxytocic activities were resolved into two peaks. The first was a fast-running component, designated "E_2." It represented the major oxytocic component of the elasmobranch pituitary, and its oxytocic effect was potentiated twofold by the addition of magnesium ions. The presence of a peak of similar mobility—although not necessarily of the same chemical constitution—has been confirmed by a number of other workers, and extended to include *Squalus acanthias* (Pacific variety = *Squalus suckleyi*), *Dasyatis sabina*, *Carcharinus leucas*, *Sphyrna mokarran*, and *Eulania milberti*. Perks (1966) has shown that the behaviour of this component is parallel to that of oxytocin in a large number of chromatography systems, and it would appear to be an oxytocinlike, neutral principle, or principles. However, it is not the same as oxytocin, since it is potentiated by magnesium ions, and it possesses other biological differences from the mammalian principle. In *Squalus acanthias*, this major component can be purified by column chromatography, and it appears to have a relatively high potency.

Column partition chromatography of the purified product, on a system which would separate oxytocin, 8 Ile oxytocin, and 4 Ser, 8 Ile oxytocin suggests that it is a homogenous peptide—but this has not yet been confirmed by amino acid analysis.

The second oxytocic peak resolved by Heller and his group was a slow-running component, which moved with the R_f of a basic principle, similar to arginine vasotocin or the mammalian vasopressins. It was designated "E_1." It showed a distinctive spectrum of biological activities, which included a remarkable sixfold potentiation of its oxytocic potency by the presence of magnesium ions. However, in *Negaprion brevirostris* the potentiation was closer to the twofold value of the fast-running E_2 peak. This suggested that the properties of the slow-running component could vary in different species. Moreover, there was reason to suppose that it did not exist in all extracts. Acher and his co-workers did not observe a peptide which combined the properties of the E_1 principle during their purification and analysis of skate extracts. Other groups, using the same paper chromatography methods as those of Heller and his co-workers, failed to detect a principle with the properties of the E_1 peptide. Perks (1966) and Swiatkiewicz et al. (1967), working with *Squalus acanthias* and other species, could only resolve a single fast-moving oxytocic peak with a uniform twofold magnesium potentiation throughout its length. At first, overloading artifacts were avoided by the use of relatively low doses of crude extracts, but later, extracts were subjected to gentle preliminary purification by gel filtration on Sephadex columns, and it was found possible to chromatograph quantities as high as 7000 mU of oxytocic activity. Even at such high levels, no E_1-type oxytocic principle could be separated on the chromatograms. In 1967, W. H. Sawyer carried out essentially similar experiments on *Squalus acanthias* and found that the great majority of the oxytocic activity ran as a fast-moving, homogeneous E_2-type peak. However, he detected a slow-running moiety, present in trace amounts, which corresponded to arginine vasotocin. Swiatkiewicz et al. (1967) confirmed this by demonstrating a slow-running region of high frog bladder and antidiuretic activity in their chromatograms; however, in their experiments, the level of this arginine vasotocinlike principle was never sufficient to show up as a second oxytocic peak. The presence of arginine vasotocin was not unexpected since it had been found in trace amounts (1% of the total oxytocic activity) in *Squalus acanthias* by W. H. Sawyer (1965b) and in *Raia clavata* by Acher et al. (1965a). Heller and his co-workers did not measure pressor, antidiuretic, or frog bladder activities in their slow-running E_1 peak. Their estimations of avian depressor activity could have been affected by contaminating vasodepressor substances—and these have been shown to exist in elasmobranch extracts, at least in terms of the rat blood pressure. Further, the high magnesium potentiation, which was the most notable characteristic of the E_1 peak, has been shown to be a variable and unreliable criterion for identifying elasmobranch principles. Finally, the ratio of milk-ejection: rat uterus oxytocic activity ($-Mg^{2+}$) of the slow-running E_1 peak of *Raia clavata*, which was 1.77 (1.47-2.07), is unlikely to be significantly different from the values of approximately 1.8 and 2.2 given for synthetic arginine vasotocin by Berde et al. (1962), and by W. H. Sawyer (1965b), respectively. Therefore, it is possible that the slow-running principle of Heller and his co-workers represents arginine vasotocin, perhaps contaminated by small quantities of vasodepressor substances, and by variable amounts of the major oxytocic principle, which might trail behind if any degree of overloading had occurred. However, it is still possible that the slow-moving moiety is a new peptide, which might fluctuate with the season, reproductive cycles, or other unknown factors. Further studies are needed to resolve this situation, but we can be certain that most elasmobranchs contain at least one neutral oxytocinlike principle, together with small quantities of arginine vasotocin.

The first purification and amino acid analysis of a major elasmobranch principle was that carried out by Perks and Sawyer in 1965. Extracts from *Raia ocellata* were purified by gel filtration and ion exchange chromatography on CM-Sephadex resin. The purified product appeared to have a specific oxytocic activity of 17.6 U/mg, which was lower than that of the principle later purified from *Squalus acanthias*, and lower than that of pure oxytocin (430 U/mg). The purified peptide from *Raia ocellata* gave a ratio of rabbit milk-ejection: rat uterus: antidiuretic activities of 8.5:1: 0.04. At the time, this unusually high milk-ejection potency suggested that it was a new principle, different from that found in other elasmobranch species. Amino acid analyses showed that it contained cystine, tyrosine, isoleucine, serine, asparagine, proline, glutamine, and glycine, in ratios compatible with a neurohypophysial principle. Soon afterward, Acher and his co-workers purified extracts from *Raia clavata*, *R. batis*, *R. naevus*, and *R. fullonica*. They found small amounts of arginine vasotocin, together with a major oxytocic component which contained the same amino acids as were present in the peptide from *Raia ocellata*. The principle was identified as 4 Ser, 8 Gln oxytocin. At the time, the high magnesium potentiation of this material (10-fold) appeared to differentiate it from the principle of *Raia ocellata* (2.6-fold), but this criterion has been shown to be variable and unreliable in elasmobranch work. The other biological data given for 4 Ser, 8 Gln oxytocin were insufficient to establish any possible identity between this peptide and the principle of *Raia ocellata*. However, Heinicke and Perks (1969a) have shown that the oxytocic peptide of *Raia ocellata* has the same pharmacological spectrum as that of *Raia rhina*, and W. H. Sawyer et al. (1969) have demonstrated that there is a close parallel between the biological properties of the principle of *Raia rhina* and those of synthetic 4 Ser, 8 Gln oxytocin, when they are assayed directly against one another. Therefore, it is probable that 4 Ser, 8 Gln oxytocin is the main oxytocic principle of all species of skate so far examined.

In contrast to the skate, direct assays of extracts from *Squalus acanthias* against synthetic 4 Ser, 8 Gln oxytocin have shown marked differences between the two active agents. This is particularly true of their rabbit milk-ejection and rat antidiuretic potencies. Preliminary amino acid analyses do not suggest the presence of 4 Ser, 8 Gln oxytocin, but they raise the possibility of a serine—valine containing peptide or of a mixture of such peptides. It is too early to make any clear statement on the nature of the principle of *Squalus acanthias*, but it is clear that it is different from that of the skates. This raises the interesting possibility that the skates and sharks possess different neurohypophysial peptides.

It is concluded that the elasmobranchs possess traces of arginine vasotocin, together with at least one major neutral peptide. While the neutral peptide of the skates is probably 4 Ser, 8 Gln oxytocin, that of *Squalus acanthias*, and perhaps of the sharks in general, is another, unidentified principle or principles, probably containing valine and serine, and, according to recent results, alanine in addition.

Actions of Neurohypophysial Principles in the Elasmobranchs

Little is known of the functions of neurohypophysial principles in the elasmobranchs. The few studies recorded have utilized mainly mammalian principles, which are different from those of the elasmobranchs themselves. Although there is no reason to assume that the functions of the principles bear any close relationship to those seen in higher vertebrates, most studies have been based on our knowledge of the mammalian pattern.

There is no evidence that neurohypophysial principles affect the water or salt balance of the elasmobranchs. The immersion of

Scyliorhinus caniculus in hypertonic seawater, enriched to 10% NaCl, failed to change the pituitary content of neurohypophysial peptides, as judged by oxytocic assay. However, it is possible that hypothalamic production kept pace with systemic loss, and therefore this negative observation is inconclusive. The injection of 20 mU/kg (rat uterus oxytocic activity) of a crude extract of *Raia clavata* pituitaries into four nurse sharks, *Ginglymostoma cirratum*, caused no significant changes in their body weights, and it would appear that there is no water-balance effect analogous to that seen in the frog. An antidiuretic action on the kidney is unlikely, since marine elasmobranchs do not suffer the extremes of osmotic conditions which may affect a terrestial vertebrate, and their urine production appears to be constantly and notably low (12.9 ml/kg/day, *Squalus acanthias*, 15.8 ml/kg/day, *Mustelus laevis*, 1.8 ml/kg/day, *Scyliorhinus caniculus*). Since water and osmotic balance are achieved by urea conservation, it is possible that investigations of the actions of neurohypophysial principles on urea metabolism could prove well worthwhile, but at the present time there are no data available.

The fact that neurohypophysial principles affect sodium metabolism in the lampreys and in the teleost fish might suggest that they have a similar action in the elasmobranchs. However, neither 8 lysine vasopressin, nor an arginine-lysine vasopressin mixture appears to affect the sodium-secreting rectal gland of *Squalus acanthias*. Perhaps the use of the oxytocinlike elasmobranch principles—or even oxytocin itself—would have been more rewarding. It would have been closer to the physiological situation in the elasmobranchs, and also more in keeping with our knowledge of the teleosts, where oxytocinlike neutral peptides have strong influences on salt metabolism, but 8 Lys vasopressin is ineffective.

Clearly, milk-ejection activity is meaningless in an elasmobranch. An oxytocic action would be more understandable, but no author has reported stimulation of the gravid or nongravid elasmobranch oviduct by either mammalian or elasmobranch extracts. Dreyer (1946) applied an ox pituitary extract (10 mU/ml, perfusing medium) to the isolated oviduct of *Raia erinaca*, *Squalus acanthias*, *Mustelus canin* and *Prionace glauca* and failed to obtain a response. He found that extracts from *Squalus acanthias* were equally ineffective. Similarly, Perks (1959) found that oxytocin in doses up to 40 mU/ml, or an extract equivalent to one-half of a neurointermediate lobe of *Squalus acanthias* per 62 ml of bathing medium, had no effect on the isolated oviduct of mature

Scyliorhinus caniculus nor on uterine strips from near-term, pregnant *Squalus acanthias*. Effects on other smooth muscles are not as consistently negative. Dreyer (1946) failed to find effects of pituitary extracts on the stomach or intestinal muscle of the same species as those in which he tested the oviducts. However, M. E. Sawyer (1933) utilized high doses of whole ox pituitary extract (5000 mU, injected intravenously, intact preparation; 25 mU/ml supporting medium, isolated preparation) and observed rhythmic contractions of mesenteric muscle of dogfish and skate *(Squalus acanthias*, *Raia* sp.?). Movements of the pyloric stomach were also observed.

Despite the inconclusive effects of neurohypophysial principles on smooth muscle, it appears that the peptides may be capable of producing a vasopressor effect in the elasmobranchs. A marked and prolonged rise of blood pressure, with inhibition of heart contractions, was found after the injection of 5000 mU, whole mammalian pituitary extract into *Raia erinacea*, *R. diaphanes*, *R. scabrata*, or *R. stabuliforis*. A similar rise of blood pressure occurred in the ventral aorta of *Squalus acanthias* after the injection of various mammalian preparations. In contrast, Waring et al. (1950) found that arginine-lysine vasopressin had little effect on the elasmobranch blood pressure, although there was a rise in the frequency and amplitude of the heartbeat. This apparent disagreement could be explained if the vasopressor effects within the elasmobranch circulation were vested in the oxytocin present in the mammalian preparations. If this were so, the oxytocinlike principles which predominate in the elasmobranch pituitary might be important pressor agents within their own species. Direct experiments are needed in this field.

Since attempts to determine the systemic effects of neurohypophysial principles have been so frequently doubtful or negative, even when extraordinarily high doses of principles have been used, it is reasonable to consider the possibility that they have only local effects within the pituitary. The histological evidence given previously suggests that it would be possible for neurosecretory materials to pass from the median eminence through the portal system to the adenohypophysis, where they might perform a regulatory function. Since each of the species examined by Meurling (1967a) possessed either a component of the portal system which reached the neurointermediate lobe or a neural lobe sinus which could facilitate the passage of substances from the pars nervosa to the pars intermedia, it appears possible that the neurohypophysis could influence the pars intermedia by a vascular route.

Besides this, nerve axons pass directly to the pars intermedia, where they form terminals on its cells has shown that damage to the preoptico-hypophysial tract of *Scyliorhinus caniculus* is always followed by melanophore dispersion and a consequent darkening of the skin. This is explained by the removal of an inhibitory action on the intermedia cells. With the removal of this inhibition, the pars intermedia cells liberate melanocyte stimulating hormone (MSH) into the circulation, and this hormone causes dispersion of pigment throughout the melanocytes of the skin. From Mellinger's experiments, it seems possible that the inhibitory influence is mediated by the neurohypophysis, and it is worth remembering that Knowles (1963, 1965a) has distinguished individual types of neurosecretory filters which pass either to the "synthetic" or to the "release" poles of the intermedia cells. Moreover, Knowles (1965b) has noted that secretomotor junctions are formed between the neurosecretory fibers and areas of the pars intermedia cells which are distinguished by a sharply delimited endoplasmic reticulum. There was some degree of correlation between the state of the reticulum and signs of hormone liberation from the neurosecretory fibers. Knowles suggests that this would be compatable with the view that a neurohypophysial principle directly affects MSH synthesis in the cytoplasm of the intermedia cell, in the region of the endoplasmic reticulum. Despite these histological indications, it cannot be certain that the inhibitory effect results from neurosecretory fibers or their peptides, since nonneurosecretory axons are also present in the preoptico-hypophysial tract. Indeed, in recent work, Meurling et al. (1969) have succeeded in preferentially cutting the neurosecretory fibers of the infundibular stem of *Raia radiata*, a species in which the neurosecretory axons are localized in a medial position. They found that the release of MSH was little affected. However, section of the lateral, possibly adrenergic nonneurosecretory axons caused profound effects in which colour responses to a white background were abolished. It is clear that further work is needed to clarity the true significance of the close relationship between the neurosecretory fibers and the intermedia cells.

Cartilaginous Fish: Holocephalians

Structure of the Neurohypophysis of the Holocephalians

This ancient group of cartilaginous fish occupies a puzzling position in the evolutionary series. Its members are of particular interest because in some ways they may be intermediate between the elasmobranchs and the bony fish.

The holocephalians have a flask-shaped pituitary which recalls that of the elasmobranchs. However, it has an individuality of its own. Fujita (1963), working in *Chimaera monstrosa*, recognized the presence of a unique and only partly formed neurohypophysial system in which the neurosecretory cells were located in the neurointermediate lobe, but this was not confirmed by Altner (1966), who found a complete preoptico-hypophysial system in the same species. It is possible that the relative absence of neurosecretory material noted by Fujita (1963) may have been a reflection of some physiological or pathological state, since Sathyanesan (1965a) and Jasinski and Gorbman (1966) have demonstrated a well-defined neurohypophysial system, rich in neurosecretion, in the related species, *Hydrolagus colliei*. The following general description of the neurohypophysis of *Hydrolagus colliei* is based on the work of these investigators.

In *Hydrolagus colliei*, the preoptic nucleus lies anterodorsal to the optic chiasma, in a location similar to that of the elasmobranchs. It is diffuse in form but highly vascularized. The cells are irregular in shape, larger than those of elasmobranchs, and not divided by size into magnocellular or parvocellular varieties, as in the teleosts. The preoptic cells contain intensely staining, aldehyde-fuchsin-positive neurosecretory material. A few cells send beaded axons toward the forebrain, but the majority give rise to two groups of notably wide axons, which unite behind the optic chiasma to form a single preoptico-hypophysial tract. This tract is rich in neurosecretion, and accumulations similar to Herring bodies are found even in the preoptic region. The tract passes in a caudal direction, and runs below the elongated nucleus

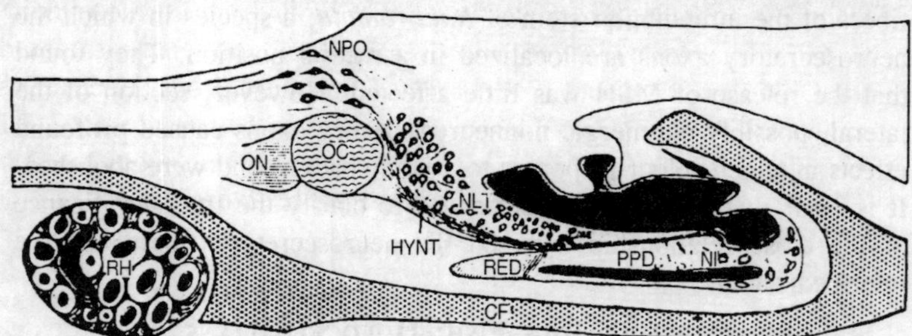

Fig. 3.4. The hypothalamus and pituitary of the Holocephalian, Hydrolagus colliei; median sagittal section from a young specimen. NPO, preoptic nucleus; HYNT, preoptico-hypophysial tract; ME, median eminence; NI, neurointermediate lobe; PPD, proximal pars distalis of adenohypophysis; RPD, rostral pars distalis of adenohypophysis; NLT, nucleus lateralis tuberis; OC, optic chiasma; ON, optic nerve; CF, cartilaginous floor of the cranium; and RH, Rachendach-hypophyse.

lateralis tuberis, which also contains aldehyde-fuchsin-positive material. In this region, the preoptico-hypophysial tract gives off ventral loops of nerve fibers, which terminate in swollen endings, full of neurosecretion, and closely applied to capillaries. These plentiful capillaries make shallow intrusions into the infundibular wall. They drain into a portal network, which is the sole blood supply to the rostral regions of the adenohypophysis ("pars distalis"). Both Sathyanesan (1965a) and Jasinski and Gorbman (1966) regard this region of the infundibular wall as a functional median eminence, where neurosecretory materials might pass to the adenohypophysis and modulate its function. No neurosecretory axons appear to cross the vascular connective tissue between the infundibular floor and the rostral adenohypophysis.

Although a few neurosecretory axons terminate in the median eminence, most continue in a caudal direction and become more densely loaded with neurosecretion. Within the large neurointermediate lobe they terminate to form a pars nervosa, which contains a rich store of neurosecretion. It is remarkable for the presence of large ganglion cells, which have been seen in both *Hydrolagus colliei* and *Chimaera monstrosa*. The pars nervosa consists of tongues of neural tissue which penetrate intimately into the pars intermedia in a manner reminiscent of such elasmobranchs as *Scyliorhinus caniculus*. The lack of connective tissue between the pars nervosa and the pars intermedia allows fine axons bearing neurosecretion to penetrate the intermedia tissue. Here, they entwine the intermedia cells, or penetrate between them and terminate as swollen endings on the wide, irregular sinusoids of the highly vascular neurointermediate lobe. These sinusoids contain blood supplied from special hypophysial arteries, which are concerned with the neurointermediate lobe in particular. The sinusoids drain into the dorsal hypophysial veins, which lead into the general circulation. The anatomical and vascular connections described by both Sathyanesan (1965a) and Jasinski and Gorbman (1966) suggest that the neurosecretory materials of *Hydrolagus colliei* could be concerned in both local control of the adenohypophysis and more general systemic functions.

In *Chimaera monstrosa*, Meurling (1967c) has shown the presence of a pituitary portal system, which connects the median eminence to the adenohypophysis, and which might allow the activity of the adenohypophysial cells to be modulated. In this species, the vascular connections of the neurointermediate lobe are well adapted to passing neurosecretory materials through the intermedia tissue into the general circulation. The neurointermediate lobe shows a close intermingling of neural and intermediate elements and possesses not only wide blood sinuses but also a neural lobe plexus, which is absent from *Hydrolagus colliei*. Although this plexus might facilitate the passage of neurohypophysial agents out of the pars nervosa, Meurling (1967c) was unable to find any accumulation of neurosecretion close to its blood vessels, as seen in the elasmobranchs *Squalus acanthias* and *Etmopterus spinax*. For this reason he considered that the neural lobe plexus did not serve for the release of neurohormones. However, he felt that this might occur more distally, so that neurosecretory materials could well have functions throughout the body.

Nature of the Neurohypophysial Principles of the Holocephalians

In 1965, W. H. Sawyer (1965b) demonstrated that crude extracts from the pituitaries of *Hydrolagus colliei* showed a pharmacological spectrum which suggested the presence of oxytocin and arginine vasotocin. This was extended by the chromatographic separation of two principles, on both carboxymethyl cellulose and carboxymethyl-Sephadex columns. The basic principle, which accounted for only 4% of the total oxytocic activity of the crude extract, had pharmacological properties similar to arginine vasotocin. The second, neutral principle was subjected to partition chromatography on G-25 Sephadex resin; it moved as a homogeneous peak, which could be distinguished from both 4 Ser, 8 Ile oxytocin and 8 Ile oxytocin. Partition chromatography did not distinguish it from the neutral principle of *Squalus acanthias*, but it showed clear differences from this principle in its pharmacological spectrum. Partition chromatography did not distinguish the *Hydrolagus* neutral peptide from oxytocin, but in this case multiple biological assays against oxytocin itself showed a close parallel between the properties of the two active agents. It must be concluded that present evidence suggests that oxytocin is the natural, neutral principle of *Hydrolagus colliei*, but no final decision will be possible until this is confirmed by chemical analysis.

The presence of arginine vasotocin in *Hydrolagus colliei* is not a surprising finding, and the low proportion of this principle is reminiscent of the situation in the elasmobranchs. However, the possible presence of oxytocin in the holocephalians is a remarkable observation, since this principle is most closely associated with the mammals and its existence so early in the vertebrate tree is not in accordance with current theories of neurohypophysial evolution. It is another example of the individuality of the holocephalians.

Actions of Neurohypophysial Principles in the Holocephalians

There is no direct evidence concerning the actions of neurohypophysial principles in the holocephalians. The most that can be suggested is that the apparent anatomical existence of a hypophysial portal system may indicate a local control of the adenohypophysis, while direct neural connections to the pars intermedia may suggest that there is a control of the intermedia cells. The direct drainage of the neurointermediate lobe sinuses into the general circulation is compatible with possible systemic functions of the holocephalian neurohypophysial principles.

BONY FISH: BRACHIOPTERYGIANS

Structure of the Neurohypophysis of the Brachiopterygians

The correct classification of the brachiopterygians (or polypterine) bony fish is disputed, but it is likely that they are the closest living relatives to the early, extinct paleoniscoid stock, and a number of authors have recognized primitive characteristics in their pituitaries. However, most attention has been paid to the open oro-hypophysial duct which connects the adenohypophysis to the mouth, and the neurohypophysial system has not received careful attention until the

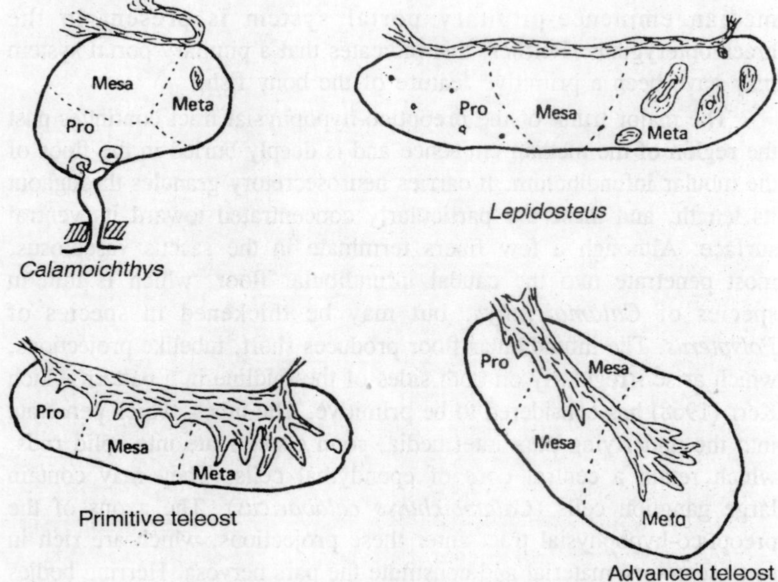

Fig. 3.5. The relationship between the pars nervosa and the adenohypophysis in various groups of bony fishes.

recent work of Kerr (1968). The following general account of the brachiopterygian neurohypophysis is based on the studies of Kerr (1968), except where the work of other authors is indicated.

The neurohypophysial systems of *Calamoichthys calabaricus*, *Polypterus senegalis*, and *P. bichir* are closely similar. The elongated preoptic nucleus is situated in a relatively rostral position and shows a weakly marked division into magnocellular and parvocellular regions. The cells contain a considerable quantity of neurosecretory granules. There are no reports that they make connections with the ventricle, but they do form a preoptico-hypophysial tract. This tract is diffuse and ill defined at first, but it becomes discrete in the floor of the hypothalamus, just anterior to the adenohypophysis. One of the most notable features of the brachiopterygians (and some chondrosteans) is a tract which divides away from the main trunk in the infundibular floor; it forms a ventral branch, which is less well defined than the main tract. This branch becomes closely associated with a plexus of blood vessels, which may penetrate relatively deeply into the ventral surface of the infundibulum. Neurosecretory granules have been seen within the palisade layer which borders these capillaries (*Polypterus* sp.). The capillaries appear to connect to the adenohypophysis and perhaps with the neurointermediate tissue; this strongly suggests that a median eminence-pituitary portal system is present in the brachiopterygians. Further, this indicates that a pituitary portal system may have been a primitive feature of the bony fish.

The major trunk of the preoptico-hypophysial tract continues past the region of the median eminence and is deeply buried in the floor of the tubular infundibulum. It carries neurosecretory granules throughout its length, and these are particularly concentrated toward its ventral surface. Although a few fibers terminate in the saccus vasculosus, most penetrate into the caudal infundibular floor, which is thin in species of *Calamoichthys*, but may be thickened in species of *Polypterus*. The infundibular floor produces short, tubelike projections, which arise irregularly on both sides of the midline in a pattern which Kerr (1968) has considered to be primitive. The tubes, which penetrate into the underlying pars intermedia, soon consolidate into solid rods, which retain a central core of ependymal cells. They may contain large ganglion cells (*Calamoichthys calabaricus*). The axons of the preoptico-hypophysial tract enter these projections, which are rich in neurosecretory material and constitute the pars nervosa. Herring bodies have been seen lining the outer surface of the neural processes, in close association with intermedia cells. The neural processes intermingle with the pars intermedia in a manner which may be more complex and intimate than that seen in the teleost pituitary (*Calamoichthys calabaricus*, *Polypterus* sp.). However, it is probable that there is little penetration into other regions of the adenohypophysis. Nevertheless, the pars intermedia, which is relatively small in some species (*Calamoichthys* sp.), does not completely invest the neural processes in all specimens of *Polypterus*, and Kerr (1968) has noted one exceptional invasion of neural tissue into the anterior regions of the adenohypophysis of this species. This intrusion could foreshadow an evolutionary trend toward the extensive neural penetration seen in the teleosts. The surface of each neural process is close to a rich capillary plexus. Some vessels pass between the neurointermediate tissue and other regions of the adenohypophysis, and they form a route by which neurosecretory materials might reach other areas of the pituitary. However, it is probable that neurohypophysial peptides could pass on into the general circulation.

Nature of the Neurohypophysial Principles of the Brachiopterygians

Studies of crude and partially purified extracts from the pituitary of *Polypterus senegalis* have shown the presence of oxytocic, avian depressor, guinea pig milk-ejection, frog bladder, and natriferic activities. The high potency in promoting water reabsorption from the frog bladder, and the level of natriferic activity, indicate the presence of arginine vasotocin. In addition, a neutral, oxytocinlike peptide has been separated by both paper chromatography and purification on carboxymethyl cellulose columns. Early pharmacological data suggested that this neutral peptide was 8 Ile oxytocin, or a mixture of this peptide with 4 Ser, 8 Ile oxytocin. However, column partition chromatography, together with a direct comparison of the neutral peptide with synthetic 8 Ile oxytocin and synthetic 4 Ser, 8 Ile oxytocin, has strongly suggested that *Polypterus senegalis* contains 4 Ser, 8 Ile oxytocin. At present there is no chemical confirmation of the structure of the peptides of this species. However, the pharmacological data suggest the presence of arginine vasotocin and 4 Ser, 8 Ile oxytocin in the primitive brachiopterygians, and this indicates that the pattern of the neurohypophysial peptides was set early in the evolution of the bony fish.

Actions of Neurohypophysial Principles in the Brachiopterygians

There is no direct information on the actions of arginine vasotocin or of 4 Ser, 8 Ile oxytocin in this group. It is only possible to point out that the anatomical relationships of the pituitary are compatible with their possible importance in the control of the adenohypophysis. They could also reach the systemic circulation, and affect tissues throughout the body.

BONY FISH: CHONDROSTEI

Structure of the Neurohypophysis of the Chondrostei

The Chondrostei, which include the sturgeons and the paddlefish, represent a distinct branch of the evolutionary tree. However, they may be closer to the main trunk of vertebrate evolution than the teleost fish. Their pituitaries are more primitive than those of the teleosts, and they are more closely allied to those of *Polypterus* and the brachiopterygians. In recent years, their neurohypophysial system has been described by Polenov and Barannikova, and the reports of these authors are the basis of the following general description.

In 1932, Charlton noted the presence of a relatively caudal and extensive preoptic nucleus in the paddlefish, *Polydon*. In the sturgeon, *Acipenser fulvenscens*, the nucleus lies in a similar position to that of the teleost fish. In *Acipenser guldenstadt* Brandt, *A. stellatus* Pallas, and *Huso huso* L. the nucleus is divided into a ventral region of small bipolar or unipolar cells, and a dorsal region of large multipolar cells. In *Acipenser fulvescens*, a number of the larger type of cells intrude among the smaller variety, so that the distinction into two zones is not as marked as in the teleost fish. The preoptic cells have single, eccentric nuclei, which are round or oval in shape. Their cytoplasm contains fine neurofibrils, Nissl substance, and clear vacuoles. More important, they contain neurosecretory granules, which are variable in size and less numerous than in the teleosts. According to Polenov and Barannikova (1958), neurosecretion is formed in the perinuclear region and later spreads throughout the cytoplasm; although granules were found around the vacuoles, these authors do not consider that they were formed by the vacuoles themselves (cf. *Scyliorhinus caniculus*). The preoptic cells extend thick dendrites between the ependymal cells which line the ventricle of the brain. They give rise also to the preoptico-hypophysial tract. This tract carries neurosecretory granules mainly in its peripheral fibers and in some specimens granules may be found throughout its length. At first, in the region of the optic chiasma, the tract consists of symmetrically arranged bundles, densely packed with secretion, and close to the surface of the infundibulum. Over the adenohypophysis, these groups of fibers merge into a single, thin, broad bundle. In *Acipenser fulvescens*, this tract gives off a ventral

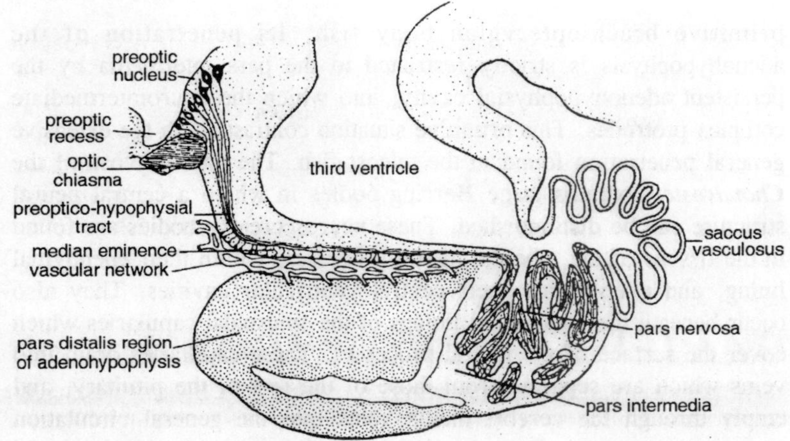

Fig. 3.6. The hypothalamus and pituitary of the sturgeons, Acipenser guldenstadt Brandt and A. stellatus Pallas.

branch, reminiscent of that seen in the brachiopterygians. This branch is a compact bundle of axons, which comes into close contact with the vascular connective tissue which lies between the infundibular wall and the adenohypophysis. Herring bodies are found in the infundibular floor. In contrast, Polenov (1966) does not describe a discrete ventral branch in *Acipenser guldenstadt* Brandt or *A. stellatus* Pallas. However, he gives a detailed description of a broad "neurosecretory contact area," where the loosely arranged, nonmedullated neurosecretory fibers of the preoptico-hypophysial tract give rise to an outer layer of fine radial axons. These axons form club-shaped, ribbonlike, or tapering terminal bulbs against the vascular connective tissue which separates the brain from the pituitary. There are no large Herring bodies in the area of contact, but the total amount of neurosecretory material is considerable because of the large size of the area involved. The connective tissue membrane becomes thinner wherever contacts are most intimate. It contains broad, sinusoidal capillaries, which penetrate the adenohypophysis in its caudal region. It is reasonable to suggest that this "contact area" is the homolog of a median eminence.

The majority of the axons of the preoptico-hypophysial tract continue caudally in the infundibular wall and spread out into a broad, thin bundle. At the end of the infundibular process, this neural lamella gives rise to large, hollow diverticuli and occasional solid processes, which form on either side of the midline. These protrusions branch and penetrate into the underlying intermedia tissue, and they constitute the pars nervosa. The pars nervosa is similar in form to that of the primitive brachiopterygian bony fish. Its penetration of the adenohypophysis is strictly restricted to the pars intermedia by the persistent adenohypophysial cavity, into which the neurointermediate complex protrudes. This primitive situation contrasts with the extensive general penetration found in the teleost fish. The pars nervosa of the *Chondrostei* contains large Herring bodies in which a central neural structure can be distinguished. These neurosecretory bodies are found in the tissues of the tubelike neural processes, within their ependymal lining, and even inside their central ventricular cavities. They also occur beneath the endothelial linings of the sinusoidal capillaries which cover the surface of the neural processes. These capillaries drain into veins which are separate from those of the rest of the pituitary, and empty through the cerebri media veins into the general circulation (*Acipenser stellatus*). It appears that neurohypophysial material could reach the rostral adenohypophysis through the primitive portal system, and it could also reach the general circulation through the sinuses of the neurointermediate lobe.

Nature of the Neurohypophysial Principles of the Chondrostei

The neurohypophysial principles of a number of Chondrosteans have received preliminary study by pharmacological methods (*Polydon spathula*, *Acipenser guldenstadti persicus* Boradin, *A. guldenstadti* Brandt, and *A. nudiventris*). The levels of oxytocic and vasopressor activity detected in the pituitaries of *Polydon spathula* suggested that they contained only arginine vasotocin, and this was supported by the failure of paper chromatography to resolve a neutral principle. However, any neutral peptide accounting for less than 6% of the total oxytocic activity would have escaped detection. This is a serious possibility, since extracts from species of *Acipenser* were subjected to paper chromatography and contained a neutral principle which never amounted to more than 3.9% of the total oxytocic activity of the major hormonal constituent, which was probably arginine vasotocin. It can be concluded that the chondrosteans so far examined appear to contain mainly arginine vasotocin, with small, perhaps variable, amounts of an unidentified neutral principle. This is the reverse situation to that seen in the elasmobranchs.

Actions of Neurohypophysial Principles in the Chondrostei

There are no direct experiments on the effects of neurohypophysial principles in the Chondrostei. Again the presence of a median eminence, and the close association of the pars nervosa with the pars intermedia, suggest that the neurohypophysis could be important in the local control of the adenohypophysis. However, anatomical considerations indicate that neurohypophysial principles could also escape into the general circulation. The only suggestion of their possible influence outside the pituitary comes from the histological studies of Barannikova and Polenov (1960). These workers noted that there were seasonal changes in the distribution of neurosecretion within the neurohypophysis. Fish in the Volga showed neurosecretion localized mainly in the preoptic nucleus in the spring and mostly in the pars nervosa during autumn. Specimens taken in the North Caspian Sea, at the point of migration and spawning, possessed a preoptic nucleus rich in vacuoles and neurosecretion. Such observations are difficult to interpret. Barannikova and Polenov (1960) suggested that the neurohypophysis was concerned in "certain vegetative functions," and they appeared to be impressed by changes in the activity of the thyroid. However, comparison with other observations in the teleosts suggests that the principles in the Chondrostei might be important in spawning or egg laying.

Bony Fish: Holostei

Structure of the Neurohypophysis of the Holostei

It is probable that the Holostei, the bowfins and the garpike, are more closely related to the teleost fish than are the other "preteleost" groups—the brachiopterygians and the Chondrostei. Therefore, it is not surprising that their general pituitary structure has been said to be similar to that of the teleost bony fish. However, this is only partly true of the neurohypophysis, for the recent work of Sathyanesan and Chavin (1967) has shown that it has clear similarities to that of the other preteleost groups. The following general account is based on the results of these authors, except where otherwise indicated.

The neurohypophysial system is well developed in the various species of bowfin, *Amia*, and of garpike, *Lepidosteus*. The preoptic nucleus is relatively short (*Amia calva*, *Lepidosteus osseus*). In *Lepidosteus osseus* and *L. platostomus* it lies antero-dorsal to the optic chiasma, on both sides of the preoptic recess, and in a position similar to that of the teleost fish. However, in *Amia calva*, the preoptic nucleus is dorsal to the pituitary; this is because the gland is drawn forward rostrally in a remarkable manner, so that it lies beneath the optic chiasma. In *Amia calva*, the preoptic nucleus closely resembles that of the teleosts in being divided into a ventral region of small cells and a dorsal region of large neurons. However, in species of *Lepidosteus*, the distinction is less clear, since some large cells of the dorsal group intrude among the smaller variety, in a manner which is reminiscent

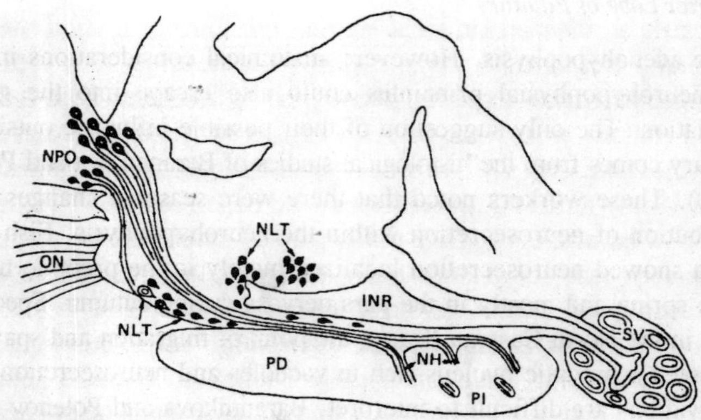

Fig. 3.7. The hypothalamus and pituitary of the garpike, Lepidosteus osseus, agittal section. ON, optic chiasma; NPO, preoptic nucleus; NLT, nucleus lateralis tuberis; INR, infundibular cavity; NH, pars nervosa; PI, pars intermedia; PD pars distalis of adenohypophysis, and SV, saccus vasculosus.

of the sturgeon, *Acipenser fulvescens*. The preoptic cells contain variable amounts of aldehyde-fuchsin-positive neurosecretion. The well-marked preoptico-hypophysial tract which leaves the nucleus can be resolved into a number of groups of fibers, which pass caudally along the infundibular floor. These fibers run below the nucleus lateralis tuberis in *Amia calva* and through it in *Lepidosteus osseus*. The nucleus lateralis tuberis is much more extensive in the holosteans than in other preteleost fish, and it may add its own fibers to the tract. These are also bound for the pituitary (*Amia calva*, *Lepidosteus osseus*). The preoptico-hypophysial tract which runs in the infundibular floor is rich in beaded droplets of neurosecretion and in Herring bodies. In *Amia calva*, the amount of stainable material is equivalent to that found in the pars nervosa—so great, in fact, that Sathyanesan and Chavin (1967) have suggested that the infundibular wall and the pars nervosa form a continuous functional unit. In *Amia calva*, *Lepidosteus osseus*, and *L. platostomus*, neurosecretory material has been seen in close association with blood vessels which penetrate the wall of the infundibulum. Frequently, Herring bodies surround these invading capillaries. The capillaries drain through the underlying connective tissue into the rostral region of the adenohypophysis. In the case of *Amia calva*, the dividing connective tissue is particularly thin, and the blood vessels which pass through it form wide sinusoids. These observations strongly suggest that the holosteans possess a simple median eminence and pituitary portal system, and this is in marked contrast to the situation in the teleost fish.

Many of the axons of the preoptico-hypophysial tract do not terminate on the infundibular blood vessels, but pass caudally, to form the pars nervosa. In *Lepidosteus osseus* and *L. platostomus*, there is no sign of a hypophysial stalk, and the ventral infundibular wall remains relatively flat throughout its length. The pars nervosa is formed by many hollow or solid neural processes, which grow downward and laterally from the caudal region of the infundibular floor. These processes penetrate the tissue of the pars intermedia, which is confined to the caudal region of the elongated pituitary. The general form is reminiscent of that seen in other preteleost groups, particularly in the sturgeons. In *Amia calva*, the general shape is different, but the essential relationships are similar. In this species, the infundibular wall folds outward to form a funnel which projects in a rostral direction below the optic chiasma and intrudes into the pars intermedia. Fingerlike processes arise from the funnel and ramify between the intermedia cells; unlike the neural intrusions of the teleosts, these processes are confined to the pars intermedia tissue. The processes of the pars nervosa contain a few large, argyrophylic ganglion cells, but they are more notable for their rich accumulation of neurosecretory granules and Herring bodies. In *Lepidosteus osseus*, the neurosecretory axons have been seen to terminate against blood vessels, presumably against those which form a plexus around the neural processes of the pars nervosa. It appears that neurosecretory materials could be liberated into these capillaries, and perhaps reach the general circulation; the presence of a pituitary portal system suggests that they might also reach the adenohypophysis.

The holostean neurohypophysis is typical of the preteleost bony fish in most of its characteristics. It is strikingly different from the teleost in its possession of a pituitary portal system. However, in a few points, such as the particularly clear division of magnocellular and parvocellular regions in the preoptic nucleus of *Amia calva*, there are hints of the evolution of the teleosts.

Nature of the Neurohypophysial Principles of the Holostei

Pharmacological and chromatographic evidence suggests that the holosteans contain the same neurohypophysial principles as those of the teleost fish. Follett and Heller (1964a) have used paper chromatography to resolve extracts from the pituitaries of *Lepidosteus platostomus* and *Amia calva* into two active componants. The slow-running principle had oxytocic, rat vasopressor, avian vasodepressor, and milk-ejection activities which were compatible with arginine vasotocin. The fast-running moiety gave oxytocic, avian depressor and milk-ejection effects which were quantitatively similar to 4 Ser, 8 Ile oxytocin. W. H. Sawyer (1966a, 1968a) investigated the peptides of the pituitaries of *Lepidosteus osseus* by both paper chromatography and column chromatography on carboxymethyl cellulose. He separated a principle with high frog bladder activity, apparently arginine vasotocin, from a neutral peptide which could not be distinguished from 4 Ser, 8 Ile oxytocin. The results of these two groups of workers confirm one another and suggest that the Holostei produce arginine vasotocin and 4 Ser, 8 Ile oxytocin. However, chemical evidence for these identities is still needed.

These results indicate that arginine vasotocin and 4 Ser, 8 Ile oxytocin were probably established as neurohypophysial peptides at an evolutionary stage before the modern holosteans diverged from the teleostean bony fish.

Actions of Neurohypophysial Principles in the Holostei

There are no direct studies of the actions of arginine vasotocin or 4 Ser, 8 Ile oxytocin in the holostean bony fish. It can only be suggested that the histological evidence suggests a possible role in the control of the adenohypophysis and that systemic effects might well parallel those found in the closely related teleosts.

Bony Fish: Teleosts

Structure of the Neurohypophysis of the Teleosts

The pioneer work of Scharrer (1928, 1930, 1932), which was carried out on teleost fish, first showed the presence of neurosecretory cells in the hypothalamus. After Palay (1945) had traced neurosecretory material throughout the neurohypophysial system of a number of teleost species, both Bargmann and Scharrer utilized Gomori's chrome-hematoxylin-phloxin stain to demonstrate the presence of a well-developed neurohypophysial system throughout a wide range of teleost fish. Since this time, the teleost pituitary has been widely studied, partly because of the easy availability of many species, and partly because of its interesting specializations. Probably the most interesting of these is the remarkable penetration of neurosecretory axons throughout all the regions of the adenohypophysis. This is not only unique to the teleost fish, but it is also in marked contrast with the situation in the mammals. Recently, the structure of the teleost neurohypophysis and its relationship to the adenohypophysis has been investigated by means of the electron microscope, and the remarkable studies of Lederis (1962, 1964), Knowles and Vollrath (1965a,b,c, 1966a,b), and Leatherland (1967) have resulted in the teleost neurohypophysis becoming one of the best understood among the lower vertebrates.

Studies of a large number of teleosts have shown that their preoptic nuclei lie approximately dorsal to the optic chiasma on both sides of the third ventricle. There are variations in both the position and the length of the nucleus in different species, with a tendency for it to become located more rostrally in more advanced species. In the relatively primitive eel, *Anguilla anguilla*, each nucleus consists of a thin sheet of neurons which extends close beneath the ependyma of the third ventricle, in the shape of an inverted "L". Charlton's diagrams (1932) suggest that this pattern is relatively common, while Oztan's description (1963) of the preoptic nucleus of *Platypoecilus maculatus* as being arc-shaped may indicate slight variations from this basic theme. The cells of the nucleus are surrounded or separated by processes from the overlying ependymal cells, and by other structures which include glial cells, extracellular colloid droplets of unknown significance,

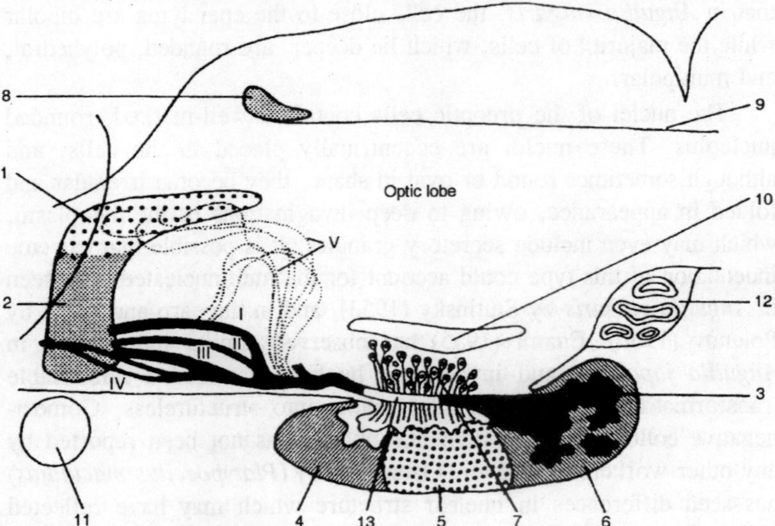

Fig. 3.8. The hypothalamus and pituitary of the teleost, Anguilla anguilla, to show the preoptico-hypophysial tract; diagrammatic representation. (1) Preoptic nucleus, pars magnocellularis; (2) preoptic nucleus, pars parvocellularis; (3) pars nervosa; (4) rostral pars distalis or pro-adenohypophysis; (5) proximal pars distalis or meso-adenohypophysis; (6) pars intermedia or meta-adenohypophysis; (7) nucleus lateralis tuberis; (8) subcommissural organ; (9) Reissner's fiber; (10) third ventricle; (11) optic chiasma; (12) saccus vasculosus; and (13) subterminal region. I-V indicate axonal bundles from the preoptic nucleus.

and axons of unknown connections (*Anguilla anguilla*). In *Salmo irideus*, Follenius (1963) has seen synapses which contact the surface of the preoptic cells, and which may well be important in their control.

Unlike the preoptic nucleus of the elasmobranchs, that of the teleosts is divided into two contiguous regions which appear to differ mainly in the size of their cells (e.g., *Porichthys notatus*). The ventral and more rostrally localized region is called the pars parvocellularis, and it consists of small, rounded, and closely packed cells, which are without any internal vacuoles in *Platypoecilus maculatus*. The dorsal region, which extends much further in a caudal direction, is termed the pars magnocellularis, and it consists of those large cells which first called attention to the neurosecretory activity of the hypothalamus (*Phoxinus laevis*, *Fundulus heteroclitus*). Its cells are loosely packed in *Anguilla anguilla*, occur in small clusters in *Carassius auratus*, but form larger groups in *Gadus morrhua*, *Cottus bubalis*, and *Salmo irideus*. Their shape has been described by a variety of terms—flask-shaped in *Carassius auratus*, fusiform or round in *Porichthys notatus*, and polygonal in *Platypoecilus maculatus*. However, Stutinsky (1953) has pointed out that in *Anguilla vulgaris*, the cells close to the ependyma are bipolar while the majority of cells, which lie deeper, are rounded, polyhedral, and multipolar.

The nuclei of the preoptic cells contain a well-marked, rounded nucleolus. These nuclei are eccentrically placed in the cells, and although sometimes round or oval in shape, they become irregular and folded in appearance, owing to deep invaginations of the cytoplasm, which may even include secretory granules. It is possible that extreme indentation of this type could account for the multinucleated cells seen in *Anguilla vulgaris* by Stutinsky (1953), and in the carp and sazan by Polenov (1960). Enami (1955) has observed similar indentations in *Anguilla japonica*, and in addition he has described a remarkable transformation of whole cellular nuclei into structureless, Gomori-negative colloid. This fundamental change has not been reported by any other workers. However, Oztan (1963) (*Platypoecilus maculatus*) has seen differences in nuclear structure which may have reflected different states in the activity of the preoptic cells, and Polenov (1960) (carp, sazan) has observed apparently pycnotic nuclei in cells which were densely loaded with neurosecretion.

Outside the nucleus, the cytoplasm of the neurosecretory cells can be divided into an inner and an outer zone. Usually, the outer, peripheral zone contains only the membranes and cisternae of the endoplasmic reticulum (Nissl substance), but electron-lucent vesicles have been seen in the eel, *Anguilla*. In contrast, the inner, central zone contains many inclusions, some of which do not fall into any ready classification. Nevertheless, rod-shaped mitochondria have been seen in many species, for example, in *Gadus morrhua*. Large, round vacuoles, or smooth surfaced vesicles with electron-lucent contents have been described in *Carassius auratus*, *Gadus morrhua*, and *Zoarces viviparus*. Scharrer (1962), working with *Opsanus tau*, has reported that similar large vesicles may have openings which expose their finely granular contents to the outer cytoplasm, but this may be an artifact of fixation. In addition, Scharrer has reported the presence of cilia in the neurosecretory cells of this species. The Golgi complex is found in the central zone, and it is well developed; it has been described as consisting of narrow tubules and vesicles, or as groups of agranular membranes, in *Gadus morrhua*, *Cottus bubalis*, *Salmo irideus*, and *Perca fluviatilis*. There is almost universal agreement that the Golgi complex produces the elementary vesicles (= elementary granules) which are the neurosecretory material, as seen under the electron microscope.

The elementary vesicles, which can occur throughout the central zone of the cytoplasm, have a complex structure. The outer surface is smooth and consists of a single membrane, and immediately within it is a narrow, clear zone. The center of the vesicle is full of a fine, osmiophylic, electron-dense, and perhaps lipid-containing granular material. These elementary vesicles, which are seen in the cytoplasm of the preoptic cells, are larger than the closely similar structures found in the neighbouring nucleus lateralis tuberis. Although their size may vary between different species, it is often possible to divide the elementary vesicles of a particular species into two distinct ranges. If these correspond to the similar vesicles present in the pars nervosa, as would seem reasonable, studies in *Gadus morrhua* would suggest that arginine vasotocin is associated with those of smaller size. Leatherland and Dodd (1967) have made the important observation that the two different sizes of elementary vesicles present in the eel, *Anguilla anguilla*, can be correlated with two different types of preoptic cell. The larger elementary vesicles, 2150 ± 30 Å in diameter, are associated with a more electron-dense type of cell, which contains swollen and irregular cisternae in its cytoplasm. In contrast, the smaller elementary vesicles, 1627 ± 31 Å across, are found in more electron-lucent preoptic cells, which possess a narrow and regular endoplasmic reticulum. These results lead to the interesting speculation that the two neurohypophysial hormones of the teleost fish may well be made by different cells within the preoptic nucleus.

Studies with the classic staining techniques suggest that neurosecretory materials, once formed, concentrate around the nucleus of the cell, and then spread out into its processes. In some species, the secretion may leave the cell in greater quantities at particular seasons, perhaps in correlation with the breeding cycles; such seasonal variations in stainable neurosecretion, and in the cytology of the preoptic cells, have been reported in the carp, the sazan, in *Salvelinus leucomaenis pluvius*, and in *Anguilla anguilla*. Oztan (1966) has noted reduced numbers of elementary vesicles and smooth-surfaced vesicles in the preoptic cells of *Zoarces viviparus* during summer, and also in response to continuous light.

Neurosecretory products may leave the preoptic nucleus by a number of possible routes. Since the preoptic cells are intimately associated with a rich capillary plexus, the possibility of a direct passage into the blood of the hypothalamus cannot be discounted at present. Indeed, there are observations which suggest that it could occur. Oztan (1963) has observed that dendrites from the preoptic cells connect to nearby capillaries in *Platypoecilus maculatus*, and Sathyanesan (1965b) has noted that section of the preoptico-hypophysial tract, followed by "activation" of the system by continuous light, will produce dilation of blood vessels around the preoptic nucleus of *Porichthys notatus*. However, in general, this direct route into the hypothalamic blood vessels is not considered to be important.

In some species, neurosecretory materials may pass to the cerebrospinal fluid. In a number of species of *Anguilla*, it is clear that thick dendrites or even whole preoptic cells may protrude into the third ventricle, where they are sometimes associated with a coagulum. Enami (1954) has suggested that the dendrites continue as Reissner's fiber and carry neurosecretory granules to the subcommisural organ for storage; however, these interesting possibilities have not yet been confirmed, and Leatherland and Dodd (1968) have demonstrated staining differences between the Gomori-positive materials of the preoptic nucleus and those of the subcommisural organ. Dendrites—sometimes with granules—have been seen to reach the ependyma and ventricle of the carp, the sazan, and the goldfish, *Carassius auratus*. However, this may not be a universal situation in the teleosts, for Oztan (1963) has remarked on the absence of such connections in *Platypoecilus maculatus*. This observation would prevent the general acceptance of

Knowles and Vollrath's suggestions (1965b) that the ventricular dendrites are concerned in feedback mechanisms which control the preoptic nucleus. Nevertheless, such connections are well marked in a number of teleosts, and it is reasonable to suggest that they are concerned either with sensory functions or with the conduction of neurosecretory products to the ventricle. Despite such alternative routes, most workers consider that neurosecretory materials leave the preoptic cells down the main axons of the preoptico-hypophysial tract.

The preoptic cells give rise to broad hillocks, which contain all the structures of the main cell body, and in addition, long, thin, straight canaliculi which have led Palay (*Carassius auratus*) to suggest that the axons which they produce are, in fact, enlarged dendrites. These fleshy, undulating, unmedullated axons form the preoptico-hypophysial tract (*Gadus morrhua*). They pass out of the nucleus in an irregular and diffuse manner, often leaving in a lateral direction and then bending ventrally before forming a discrete tract in the infundibular floor. In the stickleback, *Gasterosteus aculeatus*, there is an extreme condition in which the axons cross much of the infundibular floor as separate, lateral tracts, and only come together to enter the pituitary itself. The tract has been particularly clearly mapped in the eel, *Anguilla anguilla*, by Leatherland et al., 1966. Here, the pars parvocellularis gives off four well-defined and ventrally located tracts, which turn and run in a caudal direction for some distance, before they are joined by a diffuse curtain of axons which descends from the large, dorsally located cells of the pars magnocellularis; all these axons unite to form a single tract which converges on the midline, above the optic chiasma, and continues either past or through the nucleus lateralis tuberis toward the pituitary.

The main preoptico-hypophysial tract of many teleosts has been shown to carry rich accumulations of neurosecretory material, often in a beaded form. Oztan (1963) has remarked on the presence of Herring bodies in the tract of *Platypoecilus maculatus*, and she has added the interesting observation that there is an increase in the neurosecretion present in the tract of sterile fish produced as a backcross between this species and *Xiphophorus helleri*. Follenius (1963) has examined the neurosecretory droplets present in the tract of *Salmo irideus* by electron microscopy, and he has shown that they consist of "packets" of elementary vesicles, essentially similar to those present in the cell body, but now grouped into discrete masses. There is no information concerning the mechanism of formation of these masses. Follenius has shown that the elementary vesicles do not increase in size as they move down the tract, and he has concluded that there is no extra synthesis by the axons themselves. In some species, such as *Salmo irideus*, similar neurosecretory elementary vesicles have been seen in the nearby nucleus lateralis tuberis. This nucleus, which is close to the preoptico-hypophysial tract, and which is sometimes penetrated by it, also connects to the pituitary. It shows marked seasonal activity. However, its vesicles are smaller and have different staining properties from those of the preoptic nucleus; the two types of vesicle are almost certainly biologically dissimilar.

Although the main preoptico-hypophysial tract is often rich in neurosecretory granules in its more caudal regions, Leatherland et al. (1966) and Leatherland (1967) have observed that the tract of the eel, *Anguilla anguilla*, shows a small caudal accumulation of neurosecretion, which is followed by a short empty area. This area, which was termed the "subterminal area," lies above the anterior margin of the adenohypophysis, and it may represent the terminations of certain neurosecretory axons. It was not possible to relate any nerve terminations to blood vessels, but it is known that a small number of capillaries unite the vascular network of the brain with the pituitary in this species. However, these capillaries may connect to the pars nervosa. Similar scanty capillary connections between the brain and the pituitary have been noted in *Lucioperca lucioperca* and in *Salvelinus fontinalis*, and it may be pointed out that even rare capillary connections could be important if their transmitter materials were present in high local concentrations and if they were precisely directed. However, it is only in *Phoxinus phoxinus* that there has been any claim to demonstrate definite portal vessels, with Gomori-positive granules associated with their hypothalamic capillary network. In contrast, several groups of workers have felt that the teleosts lack a conventional pituitary portal system. Henderson's recent studies (1969) of the vascular supply of the pituitary of *Salvelinus fontinalis* have emphasized the possibility that in teleosts the portal vessels are replaced by a capillary network which is completely internal to the pituitary. This capillary bed joins the rostral region of the pars nervosa directly to the adenohypophysis, and although it is anatomically different from a classic portal system, it may well be its functional equivalent. It can only be concluded that the conventional pituitary portal system of the teleosts is at least anatomically underdeveloped and that it may even be absent. In connection with these problems, it is interesting to note that Sathyanesan (1965b) has shown that hypophysectomy of *Porichthys notatus* resulted in the development of a new relationship between the cut stump of the infundibulum and the local blood vessels. This was accompanied by an accumulation of neurosecretory material at the central end of the severed axons, and a similar accumulation has been seen after hypophysectomy in *Anguilla anguilla* and in *Lophius piscatorius*.

The connection of the preoptico-hypophysial tract to the pituitary is notably different in *Anguilla anguilla* and in *Lophius piscatorius*. The pituitary of *Anguilla*, and of other relatively primitive teleost fish, is closely applied to the infundibulum, in a way which is reminiscent of the preteleost groups. On the other hand, the pituitary of *Lophius*, and of many other more advanced teleosts, is suspended from the brain by a long stalk, which contains the distal preoptico-hypophysial tract. *Lophius* represents an extreme case, since its pituitary is not merely suspended and rotated, as in *Esox lucius* and *Cyprinus carpio*, but it is drawn out to lie rostral to the brain. This unusual stalk has been used in an attempt to determine whether neurosecretory cells are capable of conducting nervous impulses. Potter and Loewenstein (1955) detected impulses which were similar in conduction velocity (0.5 meter/see) and in refractory period (3-10 msec) to those of unmyelinated, Type C fibers from the frog. When these investigators stimulated the hypothalamus close to the base of the peduncle, the electrical responses of the stalk were variable and delayed, which suggested the presence of synapses between neurosecretory cells and other nerve cells. These results are interesting, but they are open to slight doubt since a few nonneurosecretory fibers are known to be present in the stalk. In some cases, the stalk appears to be a hollow structure; Sathyanesan (1965b) has noted that the cavity of the third ventricle extends into the remarkably long peduncle (4-5 mm) of *Porichthys notatus*. In the case of *Platypoecilus maculatus*, the stalk is funnel-shaped, and it shows a decussation of nerve fibers at its narrowest point; after this the axons continue into the pars nervosa, in association with blood vessels.

The preoptico-hypophysial tract of the more primitive teleosts, such as *Anguilla*, enters the pituitary gland medially, and toward its caudal region. It forms a dorsal area of neural tissue, which constitutes the most discrete part of the pars nervosa. This pars nervosa is essentially a thickening of the infundibular wall from which processes project ventrally; in structure, it is reminiscent of that of the preteleost bony fish. In more advanced teleosts such as *Lophius piscatorius* or *Cyprinus carpio*, the relationships may appear to be altered by variations in the direction of the incoming stalk, and by the tendency of the adenohypophysis to envelop the neural component, so that the pars nervosa becomes the central core of the gland. However, the pars nervosa is often hard to delineate precisely in the teleosts: Early workers realized that the adult teleost pituitary had no clearly separate lobes, and that the various regions were fused into a complex, which, although relatively consistent throughout the group, was unlike that found in the higher vertebrates. Although there is a clear division between the neural and the adenohypophysial tissues in juvenile teleosts, one of the most notable features of the adult pituitary is the penetration of the neural component into all parts of the adenohypophysis. In adult teleosts, the pars nervosa forms digitate processes which branch and form rootlike structures that penetrate deeply into the adenohypophysial tissue. In the eel, *Anguilla anguilla*, the pars nervosa retains primitive characteristics, for its caudally directed neural processes are penetrated by narrow canals which extend down from the third ventricle in a situation reminiscent of the wider canals seen in preteleost groups. In a contrasting situation found in adult specimens of *Hilsa ilisha*, Sathyanesan (1963) has described processes of the pars nervosa which pass between adenohypophysial ducts; these unusual ducts connect to a patent oro-hypophysial canal.

In accordance with the early suggestions of Diepen, Leray and Stahl (1961) divided the pars nervosa of *Mugil cephalus* into two main regions—a rostral region which penetrated the *meso*-adenohypophysis, and a caudal region which ramified intimately into the pars intermedia. The rostral region contained less neurosecretory material than the caudal division. It passed through the dorsal areas of the pituitary, close to blood vessels and came into association with acidophyl cells of the *meso*-adenohypophysis. The situation in *Hippocampus guttulatus* was similar, except that the fibers passed to basophylic, gonadotropic cells present in the same region of this species. Similar processes, characterized by the absence or near absence of stainable droplets of neurosecretion, have been seen in other species such as *Gasterosteus aculeatus*, *Perca fluviatilis*, *Lebistes reticulatus*, and *Anguilla anguilla*. Stutinsky (1953) has suggested that in the eel, *A. vulgaris*, these less heavily stained tracts may have originated partly in the nucleus lateralis tuberis, and recently this suggestion has received support from the electron microscope studies of Knowles and Vollrath, (1965c, 1966b). Knowles and Vollrath found that in *A. anguilla*, the greater proportion of the nerve fibers which invaded the *pro-* and *meso*-adenohypophysis did not stain by classic neurosecretory methods; but, nevertheless, they contained irregular neurosecretory vesicles, with central electron-dense granules, about 700 Å in diameter (Type B). Similar structures were found in the nucleus lateralis tuberis. The neural processes within the rostral regions of the adenohypophysis contained few pituicytes and no central canals to link them to the ventricles. They appeared to discharge neurosecretory products into the perivascular and intervascular spaces which permeated the rostral adenohypophysis, Knowles and Vollrath (1966b) suggested that these spaces represented the hypophysial portal system, while the fibers themselves corresponded to an internal median eminence. This is in agreement with the suggestions of other authors, who felt that the rostral region of the pars nervosa, as recognized by Diepen (1954) and by Leray and Stahl (1961), should be regarded as the functional equivalent of a median eminence. However, it must be remembered that Knowles and Vollrath (1965c, 1966b) also recognized that the rostral neural digitations contained a smaller number of axons with classic, Type A elementary vesicles, 1400 Å in diameter, and similar to those of the preoptic nucleus. The classic neurosecretory materials were rare in young specimens but became more abundant during maturity.

The great majority of the classic, Gomori-positive, nonmyelinated neurosecretory fibers (Type A) of the preoptico-hypophysial tract appear to enter the fingerlike processes of the caudal region of the pars nervosa. These penetrate extensively and intimately into the pars intermedia. The caudal neural processes contain a number of different structures. A few myelinated nerve fibers have been seen in *Perca fluviatilis* and *Anguilla anguilla*. Ganglion cells occur in *Salmo trutta*, *Gadus* sp., and *Mystus vittatus*. Pituicytes has been recognized in *Salmo irideus*, *Cymatogaster aggregata*, *Gasterosteus aculeatus*, *Lebistes reticulatus*, and *Anguilla anguilla*; in the last two species they appear to be localized mainly in the center of the neural digitations, where their processes isolate and envelop groups of neurosecretory fibers. The pituicytes have been particularly well described in *Anguilla anguilla*, by Knowles and Vollrath (1965a,b). Here they possess a deeply indented nucleus with cytoplasm which is either electron dense and vacuolated or electron lucent and filled with fibrilli. In this relatively primitive species, an epithelial type of pituicyte, with cilia or thin processes, lines the central ventricular canal which extends down the center of each neural process. The pituicytes extend from this central canal in a roughly radial manner and reach the outer surface of each process, where they contact the external intervascular space, which forms a boundary surrounding each neural digitation. The axons of the

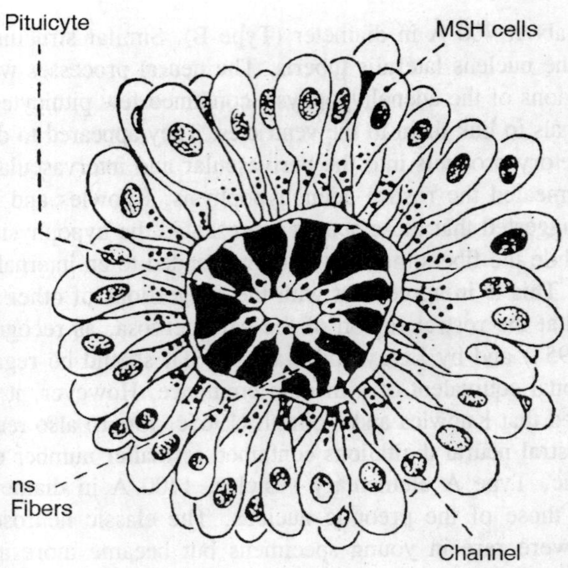

Fig. 3.9. Diagrammatic representation of the cross section of a neural process of the pars nervosa of the teleost fish, Anguilla anguilla.

preoptico-hypophysial tract pass down between the radial pituicytes, which divide them into bundles. These bundles carry droplets which stain with the classic neurosecretory stain, and the caudal processes of the pars nervosa form the chief area for the storage of neurosecretion. It is possible that species living in freshwater may store larger quantities of neurosecretion than those in seawater. In general, neurosecretory material may occur throughout the processes, or it may be densely packed toward the boundary with the pars intermedia. It occurs as irregular granules, or as beadlike droplets (*Platypoecilus maculatus*), but Herring bodies, which are present in adult fish, appear to be absent in juvenile specimens. Electron microscopy has shown that the neurosecretory material consists of elementary vesicles similar in size and form to those of the preoptic nucleus. If the results of different workers are compared, there are apparent discrepancies between the sizes of these vesicles in different parts of the neurohypophysial system, even within one species. However, these differences probably result from variations in fixation procedures, since Leatherland (1967) has noted that the sizes of elementary vesicles in the preoptic nucleus and in the pars nervosa are identical, if they are compared in the same preparation. Knowles and Vollrath (1966a), working in *Anguilla anguilla*, have suggested that the two size ranges found in these elementary vesicles indicate that the caudal pars nervosa contains two types of nerve fibers; one possesses elementary vesicles of 1600-1800 Å in diameter (Type A_1), while the second contains elementary vesicles 1200 Å across (Type A_2). It is tempting to speculate that these may correspond to the two hormonal peptides found in the teleost neurohypophysis. Lederis (1962) has noted that in *Gadus morrhua* the pressor activity typical of arginine vasotocin is associated with the relatively small elementary vesicles, 800-2000 Å in diameter, but the source of the 4 Ser, 8 Ile oxytocin has not yet been determined. Knowles and Vollrath (1966a) have recognized the presence of a few scattered fibers with small irregular vesicles, 700 Å in diameter, in the caudal neural processes; these are similar to the Type B fibers of the rostral region of the pars nervosa, so that it is probable that the fibers typical of each region are not strictly segregated from one another.

In general, the nerve fibers terminate in swellings, which appear under the light microscope either as varicose expansions of fine fibers or as greatly hypertrophied structures often termed "Herring bodies". Electron microscopy shows that the swellings contain mitochondria and large numbers of small, relatively electron-lucent vesicles, 500 Å in diameter, which are generally regarded as synaptic vesicles. In *Anguilla anguilla*, these synaptic vesicles may be clustered in the center of the fibers or they may lie along an electron-dense area next to the neural membrane. In some cases, the neural membrane is thickened at this point. These endings, which are apparently synaptic terminals, also contain elementary vesicles; and in *Anguilla anguilla* these structures appear to break down, lose their electron density, and perhaps release small particles into their surroundings.

The terminals of the neurosecretory fibers can be found in four main places: They occur on pituicytes, within the pars intermedia, on the outer intervascular channel, or around capillaries. Knowles and Vollrath (1965a,b, 1966a) have given a clear description of synapses between neurosecretory fibers and pituicytes in both *Anguilla anguilla* and *Conger conger*. The number of synapses between pituicytes and Type A_1 terminals appeared to increase when specimens of *Anguilla anguilla* were transferred to seawater, while a greater number of synapses with Type A_2 fibers were found in eels which had been placed on an illuminated white background. This suggestion of the formation of "transient" synapses between nerve fibers and pituicytes may prove to be of considerable importance.

The extent to which neurosecretory terminals penetrate directly into the pars intermedia tissue varies with different species, and it may well depend on the size of the fenestrations in the intervascular channel, which covers the surface of the digitations of the pars nervosa and corresponds to the "basement membrane" of earlier authors. In *Conger conger* there are only small areas of intervascular space which divide the pars nervosa from the pars intermedia, and one finds considerable intermingling of nerve fibers and intermedia cells, with the formation of synaptic connections. Similarly, in *Gadus* and *Phoxinus* there is little dividing structure between the two regions, and direct neuroglandular contacts have been reported. In *Lebistes reticulatus* it is probable that the dividing structures are partly developed, and a minority of the neurosecretory axons end against pars intermedia cells. In *Salmo irideus* the barrier is probably better formed, and only a few fibers pass through it, while in *Perca fluviatilis*, where the dividing structures are well developed, nerve terminations in the pars intermedia are rare. In *Anguilla anguilla* a few neurosecretory axons of type A_2 pass through gaps in the outer barrier (intervascular channel), but no synaptic terminals have been found on intermedia cells.

In *Anguilla anguilla*, by far the majority of the neurosecretory fibers (A_1, A_2, and B) terminate as a mass of rosettelike structures on the fine intervascular channel which envelops the outer surface of the neural processes. This intervascular channel is thought to correspond to the PAS-positive "basement membrane" seen by earlier optical methods. At intervals it expands to encompass capillaries within its inner space, and it would appear to be capable of transferring neurosecretory materials to either the pars intermedia cells on its outer side or to the capillaries within its framework. Jasinski (1961) has investigated the vascular supply of the pituitary of *Anguilla anguilla* and has shown that these capillaries receive their blood supply from arterioles on the dorsal surface of the pituitary; they form a fine network over the surface of the digitate processes of the pars nervosa, and although they do not penetrate the relatively avascular pars intermedia, they send ramifications into the pars nervosa tissue. Similar networks of capillaries on the surface of the neural processes, and to some extent within them, have been seen in many other species. In many teleosts these capillaries are closely associated with neurosecretory terminals, and it seems reasonable to suppose that active agents are passed into them. Follenius and Port (1962) believe that most of the blood from the pars nervosa of *Lebistes reticulatus* drains directly through the hypophysial vein into the systemic circulation, so that active materials could reach many tissues. However, Hill and Henderson (1968) have suggested that in *Salvelinus fontinalis* capillaries of the pars nervosa drain into all regions of the adenohypophysis, and that these vessels could carry neurohypophysial factors which might be important in adenohypophysial control. Recently, Henderson (1969) has greatly clarified the problem of the destination of the active agents of the neurohypophysis by a detailed analysis of the vascular supply of the pituitary of this species. She has emphasized that the pituitary is served by two largely independent arterial supplies. The first vascular route is well adapted to carry neurohypophysial products from the rostral region of the pars nervosa to the rostral and proximal divisions of the pars distalis (adenohypophysis). The second vascular circuit irrigates the caudal pars nervosa, and would appear well adapted to carry material out into the general circulation. Therefore, it is probable that active agents could be supplied both internally to the adenohypophysial cells and externally to the body tissues. However, there is no guarantee that the active agents carried in each of the two circuits are necessarily chemically similar.

It is clear that the teleost neurohypophysis has undergone considerable specialization. Two specializations appear particularly important: one is the absence, or profound reduction of the pituitary portal system, and the other is the development of direct neural contacts with the entire adenohypophysial complex. It is natural to wonder whether these notable modifications could be interrelated. Since pituitary portal systems appear to exist in more primitive vertebrates, and there are suggestions that they are present in the preteleost bony fish, it would seem possible that the development of direct nervous innervation in the teleosts has made a portal form of control obsolete. In this case, the portal system may have become almost nonexistent. It is possible that the median eminence is represented by the small subterminal area of Leatherland et al. (1966). It is also possible, and perhaps more likely, that the median eminence became incorporated into the pituitary as the rostral region of the pars nervosa. Whatever the true relationships may prove to be, anatomical considerations suggest that the neurohypophysis may control the adenohypophysis either by direct nervous action or by a diffuse liberation of active agents into its general blood supply. Similarly, it may influence the pars intermedia either by direct neural contact or by passing active agents into the intervascular channel where this exists. Finally, the neurohypophysis may pass hormones into the general circulation and control particular organs throughout the body.

Nature of the Neurohypophysial Principles of the Teleosts

As early as 1941, Heller (1941b) had demonstrated that pituitaries from *Gadus* sp. contained far more frog water-balance activity than could be accounted for by the presence of mammalian principles. In 1959, Pickering and Heller adapted paper chromatography to the study of extracts from *Gadus* sp., *Salmo* sp., and *Pollachius virens*, and succeeded in separating the "frog water-balance factor" from an oxytocinlike agent. They found that almost all the pressor and antidiuretic activity of the extract was localized with the frog water-balance factor. At the same time, W. H. Sawyer et al. (1959, 1961) showed that crude extracts from the pituitaries of *Pollachius virens* possessed remarkably strong activity on the frog bladder and on the hen oviduct. They suggested that these actions, together with the results of other biological assays, could be accounted for by a mixture of oxytocin and arginine vasotocin—a peptide which had been known only as a synthetic material up to that time. This observation was extended by Chauvet et al. (1961), who used chromatography on Amberlite IRC-50 resin to separate two active peptides from the pituitaries of *Merluccius merluccius*. Although the first principle to elute had mainly oxytocic activity, the second principle showed frog bladder, vasopressor, and natriferic activities which were consistent, once again, with the presence of arginine vasotocin. A similar separation was achieved for *Pollachius virens* by Heller and Pickering (1960, 1961); they used column chromatography on Amberlite CG-50 and obtained a frog water-balance and pressor principle which could not be distinguished from arginine vasotocin, either by chromatography or by biological assays. More important, they obtained a tentative analysis of its amino acid composition and found a general agreement with those amino acids present in arginine vasotocin. This result was extended by Rassmussen and Craig (1961); these workers used countercurrent distribution methods to purify the frog water-balance principle of *Urophycis tenuis*. They obtained a more precise analysis of the purified principle and showed that it contained the same amino acids as arginine vasotocin. They also noted that there was a general agreement between the biological potencies of their purified product and those of synthetic arginine vasotocin. Since this time, similar amino acid analyses have been obtained for *Gadus luscus*, *G. morrhua*, *Germo alalunga*, *Scomber scombrus*, and for the freshwater species, *Cyprinus carpio*. Recently, Wilson and Smith have achieved a complete sequence analysis of the frog water-balance factor of spawning salmon, *Oncorhynchus tschawytscha*, and its identity with arginine vasotocin has been clearly established.

During the early investigations, which concentrated on arginine vasotocin, the presence of an oxytocinlike peptide had been indicated in four species by pharmacological or chromatographic means. At first it was assumed that this principle was oxytocin itself since its chromatographic behaviour and its general spectrum of biological activities seemed similar to those of the mammalian principle. However, in 1961, Heller et al. made a careful comparison of the two peptides and showed that the oxytocinlike principle of *Pollachius virens* differed from oxytocin in relatively small, but significant, ways. It differed in the magnesium potentiation of its oxytocic activity, in its higher avian depressor activity, and in its lower frog water-balance potency. Further, it had almost no effect on sodium transport through the frog skin. Clearly, it was not oxytocin but a new neutral principle. In the following year, Acher et al. (1962) used chromatography on Amberlite IRC-50 resin to separate the oxytocinlike principle of *Pollachius virens*, *Merluccius merluccius*, and *Gadus luscus*. Analysis of its amino acid content, together with partial enzymic degradation, suggested that the peptide was 4 Ser, 8 Ile oxytocin. Later, the same group obtained a similar amino acid analysis for the neutral peptide of the marine species *Gadus morrhua*, *Germo alalunga*, and *Scomber scombrus*, and for the freshwater fish, *Cyprinus carpio*. The 4 Ser, 8 Ile oxytocin was synthesized by Guttman et al. (1962) and by Johl et al. (1963), and the product was found to be biologically similar to oxytocin, except for its lower intrinsic pressor and antidiuretic activities; its properties were compatible with those of the teleost principle. At the same time, W. H. Sawyer and van Dyke (1963a,b) purified the oxytocinlike principle of *Pollachius virens* and confirmed that its biological activities compared well with those of synthetic 4 Ser, 8 Ile oxytocin. Similarly, W. H. Sawyer and Pickford (1963) showed that the purified neutral principle of *Fundulus heteroclitus* was comparable to 4 Ser, 8 Ile oxytocin in its chromatographic and biological characteristics. They also found that it was preferentially lost from the pituitary of female fish during the reproductive season, while arginine vasotocin showed no seasonal changes. In 1964, Follett and Heller (1964a) confirmed, yet again, the biological and chromatographic similarity of synthetic 4 Ser, 8 Ile oxytocin and the purified oxytocinlike principle of *Pollachius virens*; in turn., they showed that the pollack peptide shared many of its properties with the neutral agents of *Salmo irideus*, *Cyprinus* sp., *Anguilla anguilla*, *Esox lucius*, and *Gadus callarias*. However, the most accurate identification of the naturally occurring 4 Ser, 8 Ile oxytocin was the complete sequence analysis achieved in spawned-out salmon, *Oncorhynchus tschawytscha*, by Wilson and Smith (1968). The successful purification and analysis of this principle was helped by the exceptionally high activities found in these specimens, which contained a total oxytocic activity of 750-1000 mU/pituitary, as compared to values which ranged from 5 to 115 mU/pituitary (*Gadus callarias*, and *Anguilla anguilla*, respectively) in the species studied by Follett and Heller (1964a).

In the mammals, it is well known that the neurohypophysial peptides are associated with a special protein, neurophysin, which probably acts as a carrier molecule within the neurohypophysis. Little is known of the neurophysinlike carriers of the lower vertebrates, but recently Pickering (1968) has isolated such a material from the pituitary of *Gadus morrhua*. This teleost neurophysin was acidic and rich in cystine, and it appeared to have a greater capacity for arginine vasotocin than for mammalian arginine vasopressin.

It can be concluded that the neurohypophyses of all the teleosts investigated so far appear to contain arginine vasotocin and 4 Ser, 8 Ile oxytocin. However, the number of species studied is minute compared with the approximately 20,000 species which exist in this highly diversified group. It is possible that isolated mutations could have resulted in the production of other active peptides, but so far none of these has been demonstrated.

Actions of the Neurohypophysial Principles in the Teleosts

Until recently, the functions of the neurohypophysial principles in the teleost fish received only sporadic attention, and the majority of studies utilized mammalian principles. Although the information available is unsatisfying, it is beginning to suggest that the peptides are concerned in at least two main fields—in osmoregulation-salt balance and in reproduction.

The earliest indications of an importance in osmoregulation came from the considerable body of histological evidence which suggested that there was a depletion of neurosecretory materials from the neurohypophysis of freshwater and marine fish when they were immersed in saline which was hypertonic to their previous environment. Lederis (1963) has shown that this is accompanied by the emptying of elementary vesicles, as seen in *Salmo irideus* with the electron microscope. Also, he carried out parallel oxytocic and pressor assays, which suggested that there was depletion of arginine vasotocin, but not of 4 Ser, 8 Ile oxytocin. This is corroborated by Carlson and Holmes (1962), who observed that the transfer of *Salmo gairdneri* from freshwater to saltwater resulted in an initial fall in the antidiuretic activity of the pituitary, while the oxytocic content tended to increase. All hormonal levels had returned to normal within 6 hr. Not all fish show these changes clearly; *Anguilla anguilla* appears to be relatively insensitive to changes in salinity, as judged by these histological criteria. The transfer or return of fish to hypotonic conditions may result in the expected accumulation of neurosecretory material in the neurohypophysis, as in *Callionymus lyra*, *Ammodytes lanceolatus*, and *Phoxinus laevis*; but *Gasterosteus aculeatus* responds with a marked fall in its neurosecretory level. Although this anomaly may be explained by possible depletion by "stress," it may emphasize the fact that the amount of neurosecretory or biologically active material present in the neurohypophysis is always a balance between the rate of hypothalamic supply and the rate of pituitary release into the circulation. Therefore, the important point made by these experiments is not the direction of the change but the fact that a change has occurred.

Despite these responses to changes in the tonicity of the environment, it is clear that the injection of neurohypophysial peptides into teleost fish does not have any water-balance effect of the type seen in the frog (*Cyprinus carpio*, *Perca flavescens*). This has been confirmed by the administration of pure arginine vasotocin (36 mμ moles/kg) to *Trichogaster trichopterus*, where there was no detectable change in body weight. Although infrequent antidiuretic effects have been seen in the teleost kidney, most attention has been focused on the strong diuretic response which usually follows injections of arginine-lysine vasopressin, oxytocin, arginine vasotocin, or teleost pituitary extracts into *Carassius auratus*. The concentration of sodium in the urine does not change, so that sodium loss is parallel to the diuresis. It is probable that this response results from an increase in the glomerular filtration rate. W. N. Holmes and McBean (1963) have shown that injections of 100 mU of either oxytocin or vasopressin will nearly double the glomerular filtration rate of *Salmo gairdneri*. Recently, this explanation has been supported by the observation that arginine vasotocin failed to cause a diuresis in the aglomerular kidney of *Opsanus tau*, even though it caused pressor effects. However, the situation may be more complicated in some circumstances, since W. N. Holmes (1959) noted that a high dose (1000 mU) of mammalian vasopressin caused a reduction in sodium output, ascribed to the kidney, in *Salmo gairdneri*, and Dlouha et al. (1967) have suggested that oxytocin causes reabsorption of sodium in the kidney of *Myoxocephalus scorpius*. Indeed, the physiological significance of the diuresis which follows the administration of neurohypophysial peptides is not clear, since it requires the use of high doses of hormones. Further, a number of workers have noted that there is a profound fall—not a rise—in glomerular filtration rate when freshwater fish enter a marine environment—and this is at a time when loss of neurosecretory material from the neurohypophysis is supposed to take place. However, the diuresis, and its concomitant sodium loss are transitory, and they are outlasted by more profound affects on sodium movement in and out of the fish at extrarenal sites such as the gills. In freshwater fish, such as *Carassius auratus*, it has been shown that oxytocin and 4 Ser, 8 Ile oxytocin stimulate influx of sodium, probably at the gills, while arginine vasotocin will enhance both influx and outflux. In the marine fish, such as *Platichthys flesus*, oxytocin increased sodium outflux, and arginine vasotocin was potent in promoting rapid sodium exchanges. Lysine vasopressin had little or no potency in these tests. These results can be summarized by saying that neurohypophysial principles (except lysine vasopressin) enhanced sodium fluxes against the prevailing osmotic gradients; they accelerated the process of adaptation to an environment of higher salinity.

It is possible that diuretic responses in teleost fish result from general vascular effects of the neurohypophysial peptides. There has been little work on possible circulatory effects of these principles in teleosts. However, recently Lahlou et al. (1968) have found pressor effects in *Opsanus tau*, after the injection of low, possibly physiological doses of arginine vasotocin (5 ng/kg). They obtained a good log dose-response curve from their injections.

Maetz (1963) has pointed out the importance of avoiding shock or "stress" effects, which often cause a "laboratory" diuresis during experiments on salt-water balance in teleosts. "Stresses" such as hypoxaemia or handling are known to result in histological changes, with depletion or accumulation of neurosecretion in the neurohypophysis. Since Stevens and Randall (1968) have shown that in *Salmo gairdneri*, adrenal medullary hormones may be liberated as a physiological mechanism for the control of the circulation during moderate or strong exercise, it would be interesting to know whether neurohypophysial principles are liberated in the same circumstances. Perhaps such moderate levels of "stress" could cause the release of neurohypophysial peptides; these might help in controlling sodium fluxes across the gills during the rapid water movement which must occur during respiratory activation owing to exercise.

Adrenal cortical tissue could also be involved in conditions of "stress" and in sodium balance. It is interesting to find that Rasquin and Stoll (1957) have shown that injections of mammalian arginine-lysine vasopressin into *Astyanax mexicanus* led to a loss of glycoprotein from basophylic cells in the adenohypophysis, and they interpreted this as a possible release of adrenocorticotropic hormone (ACTH) into the circulation. However, it is not certain that the pituitary-adrenal axis is activated in all species. Another possible indication of effects on the adenohypophysis comes from suggestions that neurohypophysial peptides may cause the release of gonadotropins; Bacon and Ball have noted that mammalian pars nervosa extracts appear to cause the maturation of free-spawning minnows. However, despite the highly suggestive nature of the anatomical connections between the neurohypophysial system and the adenohypophysis, there is still little physiological evidence for any functional relationship between the two regions.

There are a number of lines of evidence which suggest a possible connection between the neurohypophysis and reproduction in teleosts. Most of the evidence comes from observations of seasonal changes in the histology of the system, and it is often assumed that reproductive cycles are the most likely correlate. Polenov (1960) has seen seasonal variations in the preoptic nucleus of the carp and sazan; however, the pars nervosa of this species was consistently full of material. Other workers have seen seasonal changes in the granular content of the neurohypophysis of *Salmo salar* and *Anguilla anguilla*, and these changes may have been correlated with migratory activity. Perhaps the most interesting correlation has been that with the process of mating or spawning. In *Salvelinus leucomaenis pluvius*, neurosecretion within the neurohypophysis increased in quantity until the time of spawning, when it was found concentrated around the blood vessels of the pars nervosa; after spawning, it had decreased in quantity. In *Fundulus heteroclitus*, Sokol (1961) noted a similar loss of neurosecretion in both the male and female specimens during the reproductive season, but W. H. Sawyer and Pickford (1963), using pharmacological methods, were able to confirm this only in the case of the females. They made the notable observation that, in the females, there was a preferential depletion of 4 Ser, 8 Ile oxytocin, with no change in the arginine vasotocin content of their glands. The significance of these results has been increased by Wilson and Smith's recent observations on specimens of Oncorhynchus *tschawytscha*, which were collected, "spent out," after spawning. Although these glands contained an unusually high hormonal activity, they also showed an unusually low proportion of 4 Ser, 8 Ile oxytocin as compared to those of other teleosts. This was true of both sexes. It suggested that 4 Ser, 8 Ile oxytocin might have been lost preferentially during spawning. Unfortunately, the glands were not compared with controls from other stages of the life cycle, and no clear conclusion can be drawn. However, the results are suggestive, when combined with other data.

The role of neurohypophysial peptides in the females has been suggested by the work of Egami and Ishii (1962); these workers found that the injection of neurohypophysial extract into gravid specimens of *Oryzias latipes* precipitated egg laying. A similar effect was obtained in *Gambusia* sp., together with behaviour typical of spawning. Extracts from fish or frog pituitaries were said to be more effective than mammalian preparations. Although egg laying also occurred in *Rhodeus ocellatus*, species such as *Misgurnus anguillicaudatus* and certain salmonids did not respond to neurohypophysial peptides, so the effect may not be universal.

A possible function for neurohypophysial principles in the male has been suggested by the work of Wilhelmi et al. (1955). It was found that injection of crude or purified teleost pituitary extracts into *Fundulus heteroclitus* produced reflex movements reminiscent of spawning behaviour, and this effect could be mimiked by high doses of synthetic oxytocin or purified arginine vasopressin (20-60 mU/g). Similar activity has been found in extracts from *Perca fluviatilis*. It is probable that the effect is mediated by the nervous system; it does not require the presence of the gonads, and it can be elicited in both sexes.

It is too early to be certain of the precise roles of the neurohypophysial peptides in the physiology of the teleosts. If the scanty shreds of evidence available are considered, it seems possible that arginine vasotocin is concerned in osmoregulation and salt-water balance, while 4 Ser, 8 Ile oxytocin has a role in reproduction, particularly in mating and egg laying. This would be an interesting parallel with the mammal, where the basic principle (arginine vasopressin) is concerned in kidney function, and the neutral principle (oxytocin) is important in reproduction. However, further work may well prove that these speculations are unfounded.

Lungfish: Dipnoi

Structure of the Neurohypophysis of the Dipnoi

The Dipnoi, or lungfish, are important because they may be considered to occupy a phyletic position between the ray-finned bony fish and the tetrapods. Although the amphibians have not descended directly from the lungfish, it is likely that they shared a common crossopterygian ancestor, and until the pituitaries of crossopterygian species become available for study, our picture of the evolution of the neurohypophysis from fish to tetrapod must depend on investigations of the lungfish. As early as 1926, de Beer was struck by the strong resemblance between the lungfish pituitary and that of the amphibians. Since this time, many workers have made similar observations; in fact, the diagrams of Wingstrand (1966) show that it is not easy to distinguish a sagittal section of the pituitary of *Protopterus*, the African lungfish, from that of a urodele amphibian. This resemblance is partly true of the neurohypophysis, which possesses a well-marked, dorsally

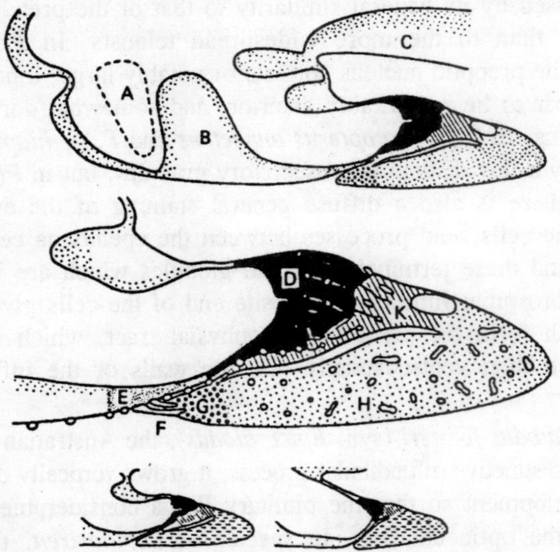

Fig. 3.10. The hypothalamus and pituitary of the lungfish, Protopterus annectens, sagittal section. Upper diagram: the relationship of the pituitary to the hypothalamus. A indicates preoptic nucleus; B, optic chiasma; and C, rhombencephalon. Middle diagram: the relationship of the pars nervosa to the adenohypophysis. D indicates pars nervosa (neural lobe); E, median eminence; F, portal blood vessels; G, "pars tuberalis"; H, pars distalis; and K, pars intermedia. Lower diagram: the relationship of the pituitary of Protopterus (left) to that of the urodele amphibian Ambystoma (right).

located pars nervosa, similar to that of an amphibian in its gross morphology. However, it retains the more intimate intermingling of neural and intermediate elements which is typical of the bony fish. Recently, Kerr and van Oordt (1966) have remarked that the intermingling of the pars nervosa and the pars intermedia in *Protopterus aethiopicus* is comparable to that of the elasmobranchs and primitive bony fish, so that the lungfish form a good transitional group between the fish and the tetrapods. The general anatomy of the lungfish pituitary has been clearly described by Wingstrand (1956), and the neurohypophysial system has been carefully studied by Dorn (1957), and by Kerr and van Oordt (1966). The recent studies of Kerr and van Oordt (1966) are the basis of the following general description of the neurohypophysial system of the Dipnoi.

The preoptic nucleus of *Neoceratodus forsteri* lies dorsal to the optic chiasma but rostral to the average position found in teleosts. It is one of the shortest nuclei found in any fish, and Charlton (1932) was impressed by its general similarity to that of the preteleost bony fish rather than to the more widespread teleosts. In *Protopterus annectens* the preoptic nucleus consists of notably large, bipolar cells, which appear to lie remarkably anterior, and somewhat dorsal to the optic chiasma. In both *Protopterus annectens* and *P. aethiopicus*, they contain only a few distinct neurosecretory granules, but in *Protopterus annectens* there is also a diffuse general staining of the cytoplasm. The preoptic cells send processes between the ependyma cells of the ventricle, and these terminate in small globules which are immersed in the cerebrospinal fluid. The opposite end of the cells gives rise to axons which form the preoptico-hypophysial tract, which runs first laterally and then ventrocaudally into the walls of the infundibular process.

Neoceratodus forsteri (syn. *Epiceratodus*), the Australian lungfish, possesses a distinctive infundibular process; it grows vertically downward during development so that the pituitary lies a considerable distance ventral to the optic chiasma. In species of *Lepidosiren*, the South American lungfish, it remains oriented in a ventrocaudal direction. In *Protopterus annectens* and *P. aethiopicus*, the African lungfish, the infundibulum is a wide, flattened, and well-delimited funnel, which projects caudally from the floor of the brain; its ventral wall is almost horizontal, and its caudal extremity is divided into two hollow pockets similar to those found in the lampreys.

In *Protopterus annectens*, Dorn (1957) has shown that the preoptic axons are spread widely as they enter the infundibular process, and there appears to be a tendency for them to be divided into different tracts. In *Protopterus aethiopicus*, the preoptico-hypophysial tract which enters the infundibulum is notably diffuse, but its clearest component curves laterally through the walls of the infundibular process, on both sides. It subdivides into upper and lower tracts in a manner reminiscent of preteleost species such as *Polypterus senegalis*, and the sturgeon, *Acipenser fulvescens*. Many of the axons of the ventral tract contact a dense capillary network, which has been seen to run in furrows within the infundibular wall, just anterior to the rostral margin of the pituitary, in *Protopterus annectens*. This region of the infundibular wall is not thickened in any way, and is not clearly demarcated from the pars nervosa. However, it can be distinguished from the neural lobe by its lower content of neurosecretion and by the presence of the invading capillaries. The capillaries connect to short portal vessels which pass into the rostral tip of the adenohypophysis. It is clear that the lungfish possess a functional median eminence and pituitary portal system.

Although a few axons of the ventral tract carry beaded neurosecretory droplets past the median eminence into the neural lobe, most of the preoptico-hypophysial fibers reach the pituitary through the dorsal branch of the tract. These dorsal axons pass through the lateral and dorsal walls of the caudal infundibular process and form a pars nervosa at its extremity.

The pars nervosa originates as an outgrowth of neural tissue in the medial walls of the two caudally directed pockets, which are formed at the caudal limit of the infundibulum. It is a large, conspicuous structure, which lies dorsal to the pars intermedia, and was once referred to as the "infundibular gland". It is made up of a mass of ramifying rods of neural tissue, and of intricate tubules which are penetrated by extensions from the infundibular cavity. The tubules are lined by ependymal cells, and these cells also form a central core to many of the solid rods of neural tissue. During the embryology of *Protopterus aethiopicus*, the neural tissue is invaded by solid strands of ependymal cells, which extend from the ventricle, and then open up to form a tubular lumen along much of their length. Dodd and Kerr (1963) have remarked that the pars nervosa of *Protopterus annectens* does not contain a great proportion of neural tissue, since so much of its structure is composed of ependymal cells, spaces, lumina of tubules, and blood vessels. Nevertheless, the pars nervosa is penetrated by the axons of the preoptico-hypophysial tract, and the fibrous matrix is heavily laden with neurosecretory material. Some axons spread into the ependyma of the dorsal wall of the infundibular cavity, where

Herring bodies have been seen close to the cerebrospinal fluid. Kerr and van Oordt (1966) have noted the presence of occasional pituicytes in the pars nervosa of *Protopterus aethiopicus*.

The outer surfaces of the neural processes and tubules are often associated with the tissue of the pars intermedia. The intermedia tissue lies ventral to them, and it is largely separated from the rest of the adenohypophysis by a hypophysial cleft—except in the case of adult specimens of *Protopterus aethiopicus*, where this cleft is partly occluded. In species of *Protopterus*, the pars intermedia is mainly composed of tubules, like the pars nervosa; the cords of intermedia cells are divided by hollow spaces and by tubelike extensions from the hypophysial cleft. The extent to which the pars nervosa penetrates and intermingles with the pars intermedia appears to vary between different species. A comparison of different authors suggests that intermingling is least marked in species of *Lepidosiren*, where the pars intermedia is particularly thin. In *Neoceratodus forsteri* the invasion of neural elements into the pars intermedia appears to be more marked. In *Protopterus annectens* and *P. dolloi*, the intimate intermingling of the two tissues recalls that typical of fish, and not the more discrete and self-contained pattern of the tetrapods. In adult specimens of *Protopterus aethiopicus* there is a highly complicated interdigitation between columns and tubules extending from both the pars nervosa and the pars intermedia.

Despite this intimate intermixing of neural and intermediate tissues, the neural processes of species of *Protopterus* are bounded and separated from the intermedia cells by connective tissue. This connective tissue barrier has been described as a thin membrane in *Protopterus annectens* and as a strong sheet in *Protopterus aethiopicus*. In *Protopterus annectens*, a few neurosecretory fibers pass through it into the pars intermedia. The connective tissue investment of the neural processes contains a dense capillary network. If the comments of different authors are correlated, it seems probable that there is a rich blood supply through these vessels, although the pars intermedia is often notably avascular. Further, if the drainage of these vessels is comparable to that of the similarly located capillaries of the bony fish, it is possible that they could supply neurosecretory materials to the systemic circulation.

Clearly, the lungfish pituitary is transitional between that of the fish and of the tetrapods. It combines an essentially amphibian pattern with a fishlike intermingling of the pars nervosa and the pars intermedia. However, it is interesting to note that it is most closely related to the pituitary of the preteleost bony fish; it shares with them a short preoptic nucleus of comparable location (some species), a median eminence supplied by a special ventral neurosecretory tract, a pituitary portal system, and large, well-developed ventricular extensions which penetrate the pars nervosa. The anatomy of the lungfish pituitary is not only a clue to the evolution of the tetrapod neurohypophysis, but it is also a reminder of the remarkable degree of specialization which has been developed by the teleost bony fish.

Nature of the Neurohypophysial Principles of the Dipnoi

The close similarities between the structure of the lungfish and amphibian pituitary are borne out by the nature of their neurohypophysial peptides. The situation in the lungfish is far from clear, but their pituitaries may contain arginine vasotocin, 8 Ile oxytocin, and perhaps oxytocin itself. Follett and Heller (1962, 1963, 1964b) subjected extracts from the pituitaries of *Protopterus aethiopicus* and *Neoceratodus forsteri* to paper chromatography and separated two active moieties. The slow-running peak (R_f approximately 0.3) showed strong vasopressor activity, and estimations of its oxytocic and natriferic potencies suggested that it was arginine vasotocin. The fast-running peak (R_f approximately 0.5-0.6) was mainly oxytocic in activity and moved in a manner similar to oxytocin, or to that of a related neutral peptide. Its pharmacological properties were clearly different from those of 4 Ser, 8 Ile oxytocin, the neutral principle typical of most bony fish. Follett and Heller (1964b) noticed that the lungfish neutral peptide was similar to oxytocin in its milk-ejection activity and in its magnesium potentiation, but the relatively high avian depressor potency indicated that the two agents were not the same. Follett and Heller (1964b) suggested that their results were compatable with the presence of 8 Ile oxytocin, a peptide which was known to occur in amphibia. However, it was possible that this peptide could be mixed with oxytocin or with some new principle. W. H. Sawyer (1965a), working with *Protopterus aethiopicus*, and using similar methods, was unable to find an unusually high avian depressor activity in the neutral principle and was inclined to ascribe the actions to oxytocin itself. Further, W. H. Sawyer (1966b) was able to separate the active principles by column chromatography on carboxymethyl cellulose resin. He confirmed the presence of a basic peptide with vasopressor, antidiuretic, and frog bladder stimulating activities consistent with arginine vasotocin. However, he found that the majority of the oxytocic activity was eluted as a large peak in the position typical of oxytocin. It could not be distinguished from this peptide.

Sawyer considered all the evidence available and concluded that probably the pituitaries of *Protopterus aethiopicus* contained oxytocin, 8 Ile oxytocin, or a mixture of the two, and that the particular combination might vary with the individual, the race, the location, the season, or the treatment of the tissues. Indeed, it was known that there were considerable variations in the hormonal content of the lungfish pituitary, since the ratio between the neutral oxytocic peptide and arginine vasotocin ranged from about 2.6 to 25.0, as judged by the oxytocic and vasopressor assays (vasopressor/oxytocic = 0.04 to 0.38). However, the oxytocic principle was always present in far greater quantities. A possible variation in the proportion of oxytocin (?) to 8 Ile oxytocin was supported by W. H. Sawyer's studies (1968a) on different batches of glands. One collection of glands appeared to differ significantly from 8 Ile oxytocin and to parallel oxytocin itself. The other three batches showed pharmacological similarities to 8 Ile oxytocin. However, the general trend has been to establish the presence of 8 Ile oxytocin more firmly, and W. H. Sawyer's most recent studies (1969) have indicated that at least some glands of *Protopterus aethiopicus* contain an overwhelming or total preponderance of 8 Ile oxytocin. Nevertheless, there is still support for the possible presence of oxytocin among other species of lungfish. In *Lepidosiren paradoxa*, thin-layer chromatography has separated a principle which appears to be arginine vasotocin from a neutral peptide with the biological properties of oxytocin. However, the authors point out that 8 Ile oxytocin might have been mixed with the oxytocin. Many of these ambiguities result from the close similarity between the properties of oxytocin and of 8 Ile oxytocin. It is possible that some individual biological assays might fail to make a clear distinction between them. For this reason, the discrepancies which occurred during assays against oxytocin standards carry the greatest weight in indicating the presence of a peptide such as 8 Ile oxytocin. Similarly, the demonstration of discrepancies between an unknown and synthetic 8 Ile oxytocin, when they were compared directly by different methods of biological assay, would be better evidence for the presence of oxytocin—or some unknown analog—in the lungfish extracts. However, many difficulties would be resolved if the neutral peptides could be separated and analyzed by chemical methods.

It appears probable that the lungfish share arginine vasotocin and 8 Ile oxytocin with the amphibia. Even if they prove to possess oxytocin, this cannot be regarded as an anomaly, since Munsick (1966) has suggested that oxytocin occurs in at least one species of amphibian, *Rana pipiens*.

Actions of Neurohypophysial Principles in the Dipnoi

Lungfish are not readily available to most investigators. Nevertheless, their important position between the bony fish and the amphibians, which appear to utilize their neurohypophysial principles in somewhat different ways, has led to a surprising amount of work on the actions of the neurohypophysial principles in the metabolism of the dipnoans. However, the work is confined almost entirely to studies of water and salt balance in the African species, *Protopterus aethiopicus*.

There is no evidence that the injection of neurohypophysial principles into the lungfish causes a water-balance effect analogous to that seen in the frog. Large doses of oxytocin, or of a mammalian vasopressin preparation (2500 mU/100 g body weight), or of pure arginine vasotocin (36 $m\mu$ moles/kg) had no effect on the total body water of free-swimming specimens of *Protopterus aethiopicus*. Although it is possible that a water-balance effect might occur in estivating specimens taken from the mud in the dry season, no studies have been made at this time. However, present information suggests that the water absorption mechanisms of the skin do not come under neurohypophysial control until the level of the amphibians has been reached, with the hagfish the only possible exception.

Although there is no general water-balance effect, neurohypophysial principles do appear to affect the kidney. In *Protopterus aethiopicus*, the injection of arginine vasotocin by either the intraperitoneal route (500 ng/kg) or by the intravenous method (10-50 ng/kg) resulted in a marked increase in the volume of the urine. Oxytocin (4000 ng/kg) caused small, inconsistent diuretic responses, but 8 Ile oxytocin (4000 ng/kg) had little or no diuretic effect when given into the peritoneal cavity. Again, it is not known whether an antidiuretic effect would be elicited during the dry season. In part, the diuretic effect could result from the increase in glomerular filtration rate which was observed, but unlike the situation in such teleosts as *Carassius auratus* other factors may be involved since the sodium concentration of the urine was also increased. It is possible that the rise in urinary sodium was caused by depression or sodium reabsorption by the kidney tubules. Arginine vasotocin (500 ng/kg) was particularly potent in increasing sodium excretion in the urine, which may rise several hundredfold under its influence; in contrast, oxytocin (4000 ng/kg) had a slight effect, and 8 Ile oxytocin (4000 ng/kg) was ineffective. When arginine vasotocin was given by the intravenous route, doses as low as 10-50 ng/kg caused loss of sodium in the urine. This partly answers the criticism that the doses required to produce natriuretic responses by the intraperitoneal route are so great that the pituitary hormone content is barely sufficient for a single response. Problems of this type have been discussed during considerations of the lampreys. Although an effect on the sodium content of the urine has been observed a number of times, it is not clear whether the principles affect sodium loss by other extrarenal routes. However, their action causes an increase in the net loss of sodium from the body. The physiological significance of this enforced sodium loss is difficult to see since it would appear to be detrimental to a totally freshwater animal. Perhaps its effects are normally in a delicate state of balance with opposing influences, similar to the antagonism seen between insulin and glucagon in the control of mammalian blood sugar. Further work is needed.

The diuretic response of the kidney to neurohypophysial peptides may be partly related to the increase in glomerular filtration rate which also occurs. W. H. Sawyer (1968c) has considered the possibility that this rise in filtration rate could be mediated by direct vascular effects. The injection of doses as low as 2.5 ng/kg of arginine vasotocin into *Protopterus aethiopicus* resulted in sharp increases in the blood pressure in the dorsal aorta. However, the pressor response was outlasted by the increase in glomerular filtration rate, so that it was unlikely that the increased filtration was totally dependent on a raised systemic blood pressure. Nevertheless, it is possible that more subtle effects on the glomerular arterioles might persist after the gross change of aortic blood pressure.

There is no information concerning the role of the neurohypophysial principles in the control of the pars intermedia or of the adenohypophysis of the lungfish; all that can be said is that the anatomical relationships are suggestive of a close association. There is no information on the possible effects of the principles on reproduction. It is tempting to speculate that the neutral peptides of the Dipnoi could be important in this sphere; as W. H. Sawyer (1966b) has pointed out, they are stored in greater quantities than arginine vasotocin, but their functions are as mysterious as those of neutral principles in the nonmammalian tetrapods and in the male mammals.

Conclusion

It is dangerous to try to generalize in a group as varied as the fish. Indeed, there are more variations, both fundamental and individual, between the different classes and species of fish than exist throughout the neurohypophyses of the tetrapods. However, a few generalizations may be useful, and they will undoubtedly provide a rich field for future critics.

The complete neurohypophysial system exists from the cyclostomes up through the vertebrate scale. Presumably, its primitive origins are lost in extinct species. In the hagfish it is indistinct, and in both the hagfish and the lampreys the pars nervosa, although present, appears to be poorly developed. The pars nervosa becomes more marked as one progresses to more advanced groups of fish. Above the evolutionary level of the cyclostomes, the pars nervosa becomes closely intermingled with the pars intermedia, although the degree of this association may be somewhat reduced in some species (e.g., *Squalus acanthias*). This notable fusion between the neural and intermediate elements persists into the lungfish, even though the Dipnoi develop the more localized, dorsal pars nervosa typical of the tetrapods. The median eminence and pituitary portal system is a common feature of the neurohypophysial system from the level of the primitive hagfish up through the vertebrate tree. In its early stages of development it appears to possess connections to the neurointermediate lobe (Myxiniformes, some Elasmobranchii). We arc faced less with the problem of how and why it developed in the vertebrates, and more with the question of why it is inconspicuous in the lampreys and absent in the teleosts. The lampreys do not have a marked portal system, but the evidence presented here favours the presence of a small neural and neurovascular connection between the preoptic nucleus and the adenohypophysis; its effects may well be aided by the large sinuses which lie between the pars nervosa and the rest of the pituitary. The teleosts have developed direct neural contacts with the adenohypophysis, and it is possible that the portal system became obsolete and was lost. However, the importance of both neural and vascular connections between the neurohypophysis and the adenohypophysis suggests that the neurohypophysis may well mediate a control over the pars intermedia and over other adenohypophysial regions as well.

Arginine vasotocin has been demonstrated in most fish which have been examined, and it is strongly suspected in most others. Although it may be present in trace quantities in cartilaginous fish, its constant presence is a remarkable example of evolutionary stability. By contrast, the neutral neurohypophysial peptides show more variation. They may be absent in the cyclostomes, although this is not well established. The elasmobranchs may contain 4 Ser, 8 Gln oxytocin and other unknown principles, but the bony fish appear to utilize 4 Ser, 8 Ile

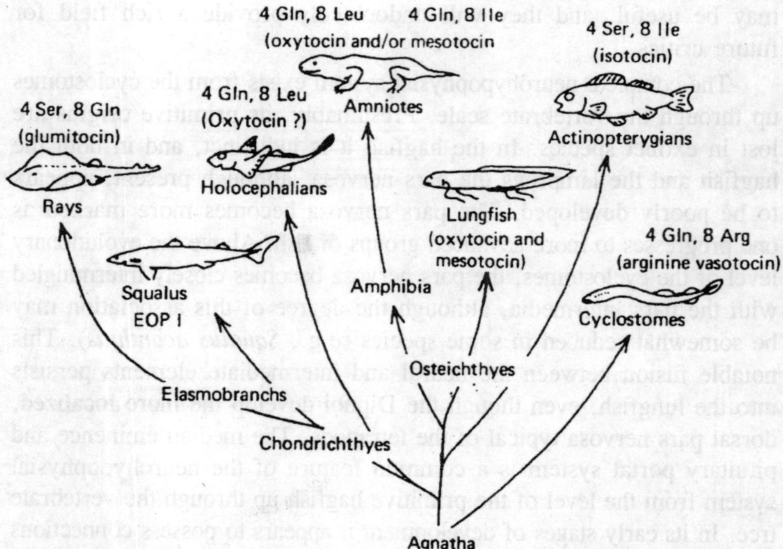

Fig. 3.11. The distribution of neutral neurohypophysial principles throughout different classes of fish.

oxytocin. The lungfish contain 8 Ile oxytocin, and perhaps oxytocin, and are effectively amphibians in terms of their neurohypophysial principles.

Although there have been clear advances in the study of the morphology of the fish pituitary, and the nature of the neurohypophysial principles is becoming better established, there is more confusion and uncertainty with regard to the actions and functions of the hormones in fish metabolism. At present there are strong suggestions that the neurohypophysial peptides are concerned in salt and water balance, most particularly on sodium metabolism. The loss of water by the kidney may be controlled by effects on glomerular filtration rate, perhaps through actions on the circulatory system. There are suggestions that the neurohypophysial peptides might be important in reproduction and influence spawning and egg laying. It is possible that the basic principles are more important in osmoregulation and salt balance, while the neutral peptides find a function in reproduction: This would parallel the situation in the mammal, but at the present time this is almost pure speculation. Although a basic pattern of function may emerge, the marked differences between the metabolism of such diverse fish as elasmobranchs and teleosts may well imply that neurohypophysial principles have widely different uses in different fish.

4

Hormones of Pituitary

In preparing a review of this sort, a contribution to a symposium in which many distinguished investigators are participating, one is handicapped by partial or complete ignorance of the ground to be covered by one's colleagues. Although this contribution is restricted, for the most part, to the pituitary hormones of fish, it is difficult to avoid all mention of topics which are treated more fully by other participants. Only two topics will be discussed: the nature of the pituitary hormones of fish, and their functions. No mention will be made of environmental influences, the role of the hypothalamus or pituitary hormonal releasing factors; and the functions of the target organs will not be discussed.

Hormones of the Fish Pituitary

As far as we know, the pituitary gland of fish is basically similar to that of mammals and contains, for the most part, similar hormones. The evidence for the presence of a particular hormone is often of an indirect nature, deduced from the results of hypophysectomy, the response to homologous mammalian hormones, or (as in the case of the growth hormone) inferred on general principles. Much of the direct evidence is derived from injections of crude fish pituitary preparations into recipients of various sorts. Negative results in such cases must be regarded with caution. For example, the presence of ACTH in crude preparations of teleostean TSH may account, in part, for the failure of such extracts to stimulate the thyroid gland of mice to any significant extent. On the other hand, all known pituitary hormones are proteins or polypeptides and, when the recipient does not belong to the same species as the donor, antibody reactions may interfere with the results, especially in the case of repeated injections. Such problems cannot be resolved until highly purified preparations are available from nonmammalian sources. The neurohypophysial hormones, for example, are known to be present in all classes of vertebrates, although weak in the elasmobranchs. Standard tests on birds or mammals have unequivocally demonstrated the presence of vasopressor and oxytocic activities in pituitary preparations from cyclostomes and teleosts. However, even within the mammals, we, know that vasopressin from cattle is chemically different from that of pigs, and we also know that there is a close structural similarity between vasopressin and oxytocin, presumably in consequence of which each of these mammalian neurohypophysial hormones contains residual activities proper to the other. The neurohypophysial hormone, or hormones, of fish and amphibians may well be chemically distinct octapeptides nevertheless capable of eliciting the standard mammalian responses for vasopressin and oxytocin. The question cannot be resolved until sufficient quantities of fish or amphibian pituitary glands can be subjected to chemical fractionation.

Intermedin, the chromatophore dispersing hormone, was one of the first to be identified in a state of partial purification from all classes of vertebrates. However, the chemical structure of mammalian intermedins is only now becoming known and new problems have arisen. In fishes, we still do not know whether the erythrophore dispersing hormone, to which the name intermedin was first applied, is identical with the fish melanophore dispersing hormone, although this seems probable, and we may well doubt whether either of these agents has the same chemical structure as alpha or beta MSH from mammalian sources. Even within the mammals there are differences in the amino acid sequence of porcine and bovine beta MSH.

In respect to the meso-adenohypophysis, the homologue of the mammalian anterior lobe, it should be stated at the outset that assignation of anterior lobe functions to this region of the fish pituitary is rarely based on direct evidence. The lobes of the fish pituitary are difficult to separate on account of the ramifications of the neurohypophysis, and most investigators have employed preparations derived from whole glands. The teleostean growth hormone, which was the first nonmammalian anterior lobe hormone to be isolated in highly purified crystalline form, provides an example. The preparation was derived from whole glands and it is only by inference, from the presence of acidophil cells in the meso-adenohypophysis, that one can assign this hormone to a specific region of the hypophysis. A few words may be said regarding the nature of the fish growth hormone. Beef growth hormone is a high molecular weight protein with two *N*-terminal groups, i.e., a forked chain; monkey growth hormone has half the molecular weight of beef growth hormone and is a straight chain with only one *N*-terminal group. Fish growth hormone has the low molecular weight of monkey growth hormone but is believed to have two *N*-terminal residues like beef growth hormone.

In respect to the other anterior lobe hormones, corticotropin has been prepared in partially purified form from salmon pituitary glands and ACTH activity is undoubtedly present in a variety of teleostean species. Its association with the meso-adenohypophysis is purely inferential. No data are available for cyclostomes, and the presumed presence of ACTH in the elasmobranch pituitary depends on indirect evidence derived from changes in the interrenal organ following hypophysectomy or administration of mammalian corticotropin.

Thyrotropin may be assumed to be present in the pituitary of cyclostomes, in view of the ability of the larval endostyle to respond to exogenous TSH and antithyroid drugs, although earlier work on the thyroids of immature adults, using ox anterior lobe extract, gave negative results. It appears from numerous investigations that the TSH content of the adult elasmobranch hypophysis is extremely low; the situation may be different in the embryo since hypophysectomy leads to an almost, total inhibition of thyroid function. On the other hand, TSH is undoubtedly present in the teleostean hypophysis. Its presumed association with the meso-adenohypophysis rests partly on indirect, but extremely convincing, evidence. Thus several investigators have reported degranulation and vacuolization of specific basophilic cells, the thyrotrophs, in this region of the teleostean pituitary under the influence of antithyroid drugs. Two Russian investigators have studied the thyrotropic potency of separate parts of the teleostean hypophysis by the induction of anuran metamorphosis. The work of Isachenko (1951) is clouded by a confusion of terminology, but her findings, if we have interpreted them correctly, appear to indicate that TSH is concentrated in the neuro-intermediate lobe of the carp—a most improbable result. However, Zaitzev (1955) has given clear evidence that TSH is associated with the meso-adenohypophysis of the pike *Esox lucius*. Phyletic specificities between crude thyrotropin preparations are well known and recent studies by Fontaine and Fontaine (1957) on the effect of temperature on the response of the thyroid of rainbow trout to mammalian and fish TSH preparations support the view that chemical differences exist. It may be noted, moreover, that Condliffe et al. (1958) have found chromatographic differences between mouse and cattle TSH.

The association of the gonadotropins with specific cell types, the gonadotrophs, in the teleostean meso-adenohypophysis rests largely on an impressive accumulation of indirect evidence. The subject has been reviewed by Atz, but special mention may be made of the findings of Sokol (1955, 1956), which demonstrated the degranulation of specific basophils in the "transitional lobe" of ovariectomized *Lebistes*. These cells were shown to be distinct from the thyrotrophs which were affected by thiourea. Kazanskii and Persov (1948) studied the gonadotropic potency of separate parts of the hypophysis of carp. The interpretation is clouded, as in the case of thyrotropin, by a confusion of terminology, but a careful study of their descriptions leads to the conclusion that gonadotropic activity was concentrated in the meso- + pro-adenohypophysis. The experiments of Barannikova (1949) on *Acipenser* leave no room for doubt that this region contains the gonadotropins. A recent cytochemical study of the carp pituitary does much to clarify the terminological confusions mentioned above and supports the interpretation which we have advanced. In the face of these findings we are confronted, however, by the work of Dodd (1955) on elasmobranchs, who claimed that in this group gonadotropic potency is associated with the neuro-intermediate lobe. This finding requires further elucidation, but it must be recalled that ACTH, which is formed in the mammalian anterior lobe, may be stored in the neuro-intermediate lobe.

It now seems certain that the teleostean hypophysis contains both an FSH-like and an LH-like gonadotropin, although the latter is present in much larger amounts. In a series of papers Otsuka (1956, a-f) has elaborated a fractionation procedure for the separation of these two gonadotropins from fish and other vertebrate sources. The latest papers give preliminary information on the amino acid composition and electrophoretic properties of the preparations.

The occurrence of prolactin in the hypophysis of lower vertebrates is still under investigation. This hormone is presumably present in the amphibian hypophysis since Grant and Grant (1958), confirming the earlier work of Chadwick (1941), have shown conclusively that purified prolactin from mammalian sources is the so-called "water drive" factor, the hormone which initiates the migration of the terrestrial red eft stage of the newt, *Triturus viridescens*, to ponds for breeding. Both older and more recent studies of fish pituitary preparations, tested by the pigeon crop assay method, have given inconclusive results. Grant is investigating the response of hypophysectomized efts to fish pituitary preparations, and, if such treatments should elicit the water drive, we may conclude that a prolactin-like hormone is present. Preliminary results, with extracts of whole glands of carp and *Fundulus*, have given positive results.

Regarding the lesser known mammalian anterior lobe hormones, an endogenous exophthalmos-producing substance (EPS) has not been demonstrated in the fish pituitary, although its presence may be inferred, at least in the teleosts. Fish are employed for the assay of EPS and a recent paper by Brunish (1958) has confirmed the fact that EPS is chemically separable from TSH. The supposed erythropoietic factor of the pituitary, which is now generally thought to be identical with ACTH, may be present in the teleostean hypophysis, since hypophysectomy leads to a state of anemia resembling that which occurs in laboratory rodents. Anna M. Slicher has shown that chronic treatment of hypophysectomized *Fundulus* with alpha corticotropin relieves the anemia of hypophysectomy.

Finally, there is present, in the hypophysis of amphibians and fishes, a melanophore-concentrating agent (MCH) which is apparently lacking in higher vertebrates. This hormone is the *W*-factor of Hogben and his co-workers. These investigators concluded that MCH is associated with the anuran pars tuberalis. Healey (1948) showed that in the minnow, *Phoxinus*, the ability to *maintain* melanophore concentration on a white background was impaired by destruction of the anterior parts of the pituitary, but not of the neuro-intermediate lobe. The pro-adenohypophysis of fishes is commonly thought to represent the pars tuberalis of mammals and the possibility that MCH is secreted by this large glandular region of otherwise unknown function deserves careful consideration. The work of Enami (1955), however, favours the presence of MCH in the meso-adenohypophysis. In clarification of a possible source of confusion in the minds of some investigators, we can state definitely that MCH is separable from MSH, as shown by preliminary assays of the two potencies in selected fish pituitary fractions.

Functions of the Pituitary Hormones of Fishes

It is obvious, from physiological considerations, that data on pituitary hormonal functions cannot be applied without reservations to fishes and amphibians. The comparative physiology of the neurohypophysial hormones provides an excellent example. We know that in amphibians, as in mammals, an important function of the neurohypophysis is concerned with the regulation of water metabolism, but the relation of the neurohypophysial hormone to the target organs has undergone evolutionary changes correlated with physiological changes in the methods of water uptake and excretion. In terrestrial anurans the neurohypophysial hormone promotes an increased uptake of water through the skin, renal antidiuresis, and water reabsorption through the bladder. These combined actions tend to conserve body water in a nonaquatic environment but, under experimental conditions, result in a marked increase in body weight following the administration of posterior lobe preparations. The situation in fishes remains obscure. There is evidence that osmotic stresses lead to the expected accumulation or discharge of neurosecretory material from the hypothalamic-neurohypophysial system in fishes, as in anurans and higher vertebrates. However, a recent paper by Arvy, Fontaine, and Gabe (1957) suggests that these changes may be mediated through changes in thyroid function. On the other hand, neither mammalian nor teleostean posterior lobe preparations appear to influence water retention either in marine or freshwater fishes. An important observation is the discovery by Sexton (1951) that pitressin promotes a state of diuresis in the goldfish, attributed to an increased uptake of water through the gills. In fishes, unlike amphibians, the skin is relatively impermeable to water and electrolytes, and the gills provide the main pathway for ionic exchange. An increased uptake of water through the gills is therefore comparable to the increased assimilation of water by the skin of an amphibian. Moreover, in purely aquatic amphibians such as *Xenopus* the dermal uptake of water, elicited by posterior lobe preparations, is not accompanied by renal antidiuresis and therefore there is an increased flow of urine. The freshwater teleost apparently responds in a strictly comparable manner, a fact which has not received general recognition. One would expect, on the other hand, that the marine teleost would possess a renal antidiuretic mechanism, but this has not as yet been demonstrated. The state of "laboratory diuresis" in captive marine teleosts suggests that such a mechanism exists and that it may be inhibited by stress. This leads to a discussion of the role of ACTH and the adrenal cortical steroids in osmoregulation, a subject that is dealt with by Dr. I. Chester Jones and his collaborators in their contribution to this symposium.

Despite our lack of knowledge of neurohypophysial functions in fishes, we know that the pituitary of teleosts is extremely rich in "neurophysine." A possible function in relation to reproduction was suggested by the accidental discovery that neurohypophysial hormones (including synthetic oxytocin) elicit the spawning reflex even in castrated or hypophysectomized *Fundulus*. Doubt has been cast on the interpretation of these results by students of fish behaviour and, as far as we know, other investigators have not been successful in attempts to stimulate sexual behaviour by this method in different species such as trout. Nevertheless, I should like to take this opportunity to answer some of our critics. We are confident that the strange behaviour observed in *Fundulus* has been correctly interpreted and is not some generalized state of convulsion elicited by abnormal doses of posterior lobe hormones. In addition to the highly characteristic S-shaped quivering, seen only in nature during spawning, we have frequently observed a cuddling of vertical objects such as the aerator tube while the spasm is in progress (in nature this would be represented by the stems of reeds where spawning normally occurs) and also, after the spasm, the fish will snap at the bottom (eating of eggs or, in their absence, an attempt to do so). None of these activities can be elicited by other types of hormonal treatment, and they do not occur spontaneously in hypophysectomized or sexually regressed fish, although readily observed during courtship and mating.

Passing now to the intermediate lobe, we have the interesting problem of a hormone (or hormones) present in all classes of vertebrates whose function is still uncertain in birds and mammals. Our knowledge of the functions of intermedin in fishes and amphibians has grown in recent years to include both morphological and physiological aspects of pigment regulation. As has long been suspected, intermedin not only promotes melanophore dispersion in reptiles, amphibians, and many species of fishes, but also (at least in teleostean fishes) stimulates new melanin pigment cell formation. This is particularly striking in a species like *Fundulus* in which injections of intermedin have no effect on the state of dispersion of the melanophores, yet chronic treatment results in the proliferation of newly formed melanocytes. Recent studies have shown that treatment with alkali greatly enhances the potency of the intermedin preparation in respect to its capacity to induce new pigment cell formation. In this connection, attention may also be called to the striking potentiation of the response by simultaneous administration of prolactin. Prolactin alone apparently promotes an increase in the amount of melanin in those melanophores which are already present but partially depleted of pigment as a result of prolonged hypophysectomy. Prolactin, unless known to be contaminated with intermedin or ACTH, has never been found to stimulate new pigment cell formation. Studies by Kosto have shown that the increase in melanin pigment is associated with a slight but significant increase in the amount of dopa oxidase in the tails of treated fish. The greatest increase was observed in those which received the combined treatment. On the other hand, Purves and Denstedt (1957) were unable to detect any increase in the amount of tyrosine or tyrosinase in the skin of intermedin-treated frogs; alkali potentiation or simultaneous administration of prolactin we presume might have yielded positive results. Approaching the problem from another angle, Bagnara and Needleman (1958) have shown that intermedin, which causes melanophore dispersion and guanophore concentration in tadpoles, leads to an inhibition of guanine formation and stimulates the synthesis of fluorescent pterins in the skin.

The presence of a normal complement of anterior lobe hormones in the meso-adenohypophysis, at least of teleostean fishes, has already been noted. The natural functions of these hormones are, in general, the same as in higher vertebrates, but some points of special interest deserve mention. We have been particularly interested in the growth hormone and its functions in teleostean fishes. It is now well established that hypophysectomized fish do not grow in length, although weight changes are sometimes confusing. Conversely, we know that both trout and *Fundulus* respond to repeated injections of purified beef growth hormone. We have recently studied the relation of this response to temperature and dose. The temperature effect shows a threshold at about 15°C and a rapid rise to an optimum plateau between 20 and 25°C. An interesting collateral effect, in this experiment, resulted from the fact that the beef growth hormone contained a small amount of LH and TSH. Both these tropins, unlike the growth hormone, showed a greater effect at low than at higher temperatures. At the time we attributed this to the inhibition of antibody release which is known to occur in frogs and fishes at low temperatures. But recently Dr. Robert W. Harrington, Jr. has shown that in *Fundulus confluentus* the gonads are stimulated by cold during the first period of growth when they are little, if at all, affected by increased illumination. This finding throws new light on the problem, since it is possible that cold favours the gonadal response rather than promotes this effect indirectly through a stimulation of the pituitary release of gonadotropin. It may be necessary to apply similar considerations to the response of the poikilotherm thyroid at low temperature.

The log dose-response curve provides a straight line relationship at doses of standard beef growth hormone varying from 3 to 30 micrograms per gram weight per injection, but the absolute magnitude varies in different experiments.

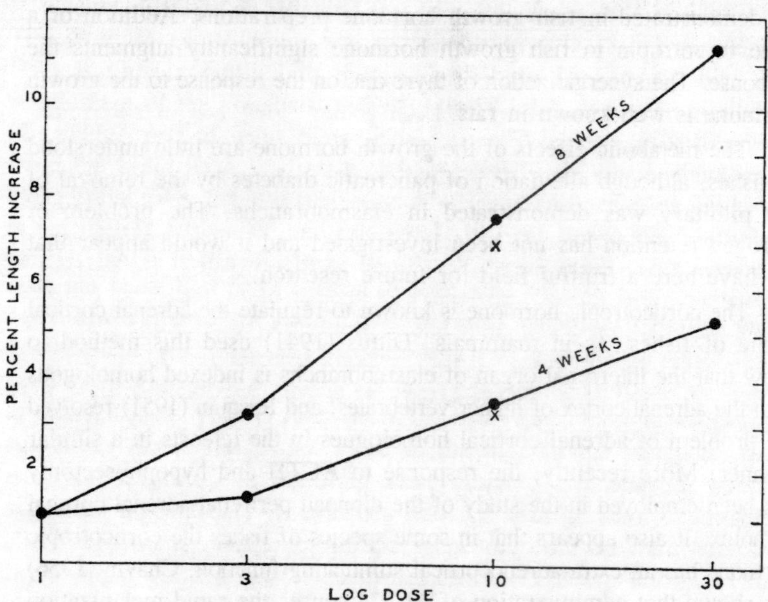

Fig. 4.1. The relationship between mean percent length increment and the logarithm of the dose of standard beef growth hormone.

Reference has already been made to the chemical differences between growth hormone preparations from different sources, but there seems to be little intelligible correlation between this and the ability, or inability, of various recipients to respond to these different preparations. Thus, *Fundulus* responds to beef growth hormone, and to fish growth hormone derived from pollack or hake, but we have recently found only a poor response to purified primate growth hormone preparations whose potency, assayed on rats, was similar to that of standard beef growth hormone. Although rats respond to monkey growth hormone the primates do not respond to beef growth hormone and rats are by no means a universal recipient since they are not responsive to fish growth hormone. These findings, suggest that phyletic relationships play little part in the problem of species specificities. A similar situation exists in respect to the gonadotropins. Moreover, even the most highly purified fish growth hormone preparations have proved to be less potent than corresponding doses of standard beef growth hormone when tested on hypophysectomized *Fundulus*. The explanation for this may reside, in part, in the fact that beef growth hormone preparations are rarely free from traces of TSH whereas this hormone could never be demonstrated in fish growth hormone preparations. Addition of a little thyrotropin to fish growth hormone significantly augments the response. The synergic action of thyroxine on the response to the growth hormone is well known in rats.

The metabolic effects of the growth hormone are little understood in fishes, although alleviation of pancreatic diabetes by the removal of the pituitary was demonstrated in elasmobranchs. The problem of nitrogen retention has not been investigated and it would appear that we have here a fruitful field for future research.

The corticotropic hormone is known to regulate the adrenal cortical tissue of fishes, as in mammals. Dittus (1941) used this method to show that the interrenal organ of elasmobranchs is indexed homologous with the adrenal cortex of higher vertebrates, and Rasquin (1951) resolved the problem of adrenal cortical homologues in the teleosts in a similar manner. More recently, the response to ACTH and hypophysectomy has been employed in the study of the dipnoan perirenal-adrenal cortical complex. It also appears that in some species of fishes the corticotropic hormone has an extra-adrenocortical stimulating function. Chavin (1956) has shown that administration of ACTH causes the rapid melanization of xanthic goldfish, although cortical steroids and, more surprisingly, intermedin have no such effect. We found that in *Fundulus* ACTH stimulated new melanophore proliferation, especially when administered in conjunction with prolactin, but in this case the response was commensurate with the inherent melanocyte stimulating activity of the corticotropin molecule and merely simulated the action of intermedin itself.

Thyrotropin, like corticotropin, regulates the functional state of the target organ in lower as in higher vertebrates, and the physiological effects of hypo- or hyperthyroidism involve poorly understood functions of the thyroid gland in fishes, a subject that is beyond the scope of the present discussion.

The gonadotropins present a difficult problem. It seems possible, from a part of the literature, that maturation of the gonads in fishes, as in higher vertebrates, may be regulated by the proper timing and interaction of at least two pituitary hormones. This would be in accordance with the evidence, cited above, that both an FSH-like and an LH-like gonadotropin are present in the teleostean hypophysis. However, experiments on hypophysectomized male *Fundulus* have not confirmed this supposition. This species is extremely sensitive to minute traces of mammalian LH and unresponsive to FSH, except in so far as the preparation contains contaminating traces of LH causing a response at high dosage levels. The regressed testes progress, under the influence of LH, up to a certain point, but we have evidence which indicates that complete maturation of the sexual products, with flowing sperm, can occur under the influence of a growth hormone preparation containing contaminating traces of both LH and TSH. In a recent study Kirshenblat (1956) has concluded that LH alone is responsible for the growth and maturation of the gonads in fishes.

The problem of fish prolactin has already been mentioned. In the past it seemed unlikely that such an agent should be present in the hypophysis of lower vertebrates, but our increasing knowledge of the synergistic role of prolactin in regulating the action of the sex hormones and gonadotropins, and the establishment of its specific role in eliciting the water drive response in red efts, throws new light on the problem. The apparent synergism between prolactin and intermedin in the development of supplementary melanophores in *Fundulus*, noted above, suggests that it may perform a natural function in the regulation of pigmentation.

In respect to the melanophore concentrating hormone it need only be pointed out that, as in the case of intermedin, the natural function in the majority of teleosts is undoubtedly the maintenance of a state of background chromatophore adaptation, since nervous impulses can only effect a transitory response.

5

THYROID GLAND

In recent years the attention of biochemists has been focused on three separate aspects of thyroid hormonal physiology: first, the nature of the hormones produced in and secreted by the gland; second, the nature of the hormones in the circulating blood plasma; and third, the nature of the hormone which acts in the peripheral tissues. Within the thyroid the attention of investigators has shifted from monoiodohistidine, mono- and diiodotyrosine, and thyroxine to the other thyronines which are produced in much smaller proportion. To date, thyroxine, 3:5:3´ triiodothyronine, 3:3´:5´ triiodothyronine, and 3:3´ diiodothyronine have been demonstrated within the thyroid, and all of these, as well as several unknowns, in the plasma. Thyroxine and 3:5:3´ triiodothyronine are the only two of the secreted thyronines definitely known to have biological activity. The others occur in minute proportions, if at all, and are of questionable activity.

The complex of hormones deriving from the gland has proved easier to identify than the form which is active in the peripheral tissues. Because of its surprisingly high biological activity, 3:5:3´ triiodothyronine has been suggested as the final form of the hormone but some recent data contradict this possibility. It is, of course, conceivable that the peripheral hormone is not one—but a group of substances including thyroxine, triiodothyronine, and their deaminated metabolites. So far, however, the deaminated analogues, such as triiodothyroacetic acid have been demonstrated only in peripheral tissues and only after the administration of radioactive thyroxine: they were never found after injection of radioiodide. Thus, the picture of thyroidal hormone synthesis which seemed so clear a few years ago has been obscured, as much as clarified, by the discovery of the numerous hormonal variants which, though small in proportion, are of high physiological activity and unevaluated significance.

As might be expected, most of the information that has been gathered concerning the nature of the thyroid hormones was derived from work with adult mammals, in particular the rat, although the high potency of 3:5:3′ triiodothyronine was well demonstrated in the tadpole metamorphosis assay. It now becomes important to learn the phyletic distribution of thyroid hormones and to recognize possible ontogenetic and phylogenetic progressions of change in hormone synthesis and hormone function in embryos and in lower vertebrates. The first question to present itself is this: is thyroxine a characteristic hormone of all vertebrate thyroids, or do different vertebrate groups utilize different ones of the iodinated amino acids? Until recently, before most of the circulating thyroid hormones were known, there seemed to be no reason to suspect that some thyroids might make a different hormone from others. Now, however, this possibility has been advanced by several kinds of experimental evidence. Smith and Matthews (1948) have shown a metabolic response in the fish *Bathystoma* after treatment with thyroids of the parrot fish. Thyroxine treatment has failed to elicit a metabolic response in any one of the many species of fish in which it has been tried. This is at least suggestive that the thyroid in fishes manufactures a hormone other than thyroxine. A similar conclusion might appear suggested by the relatively extreme sensitivity of the frog tadpole to triiodothyropropionic acid. This possible metabolite of thyroxine is 300 times as potent as thyroxine itself in inducing metamorphosis in tadpoles, *Rana pipiens*.

In order to explore the possibility of the existence of specific evolutionary patterns of thyroidal hormone synthesis, a large number of species must be examined. Fortunately, sufficient analyses of thyroxinogenesis have now been made to permit some generalizations about the most abundant iodinated compounds (iodides, monoiodotyrosine, diiodotyrosine, 3:5:3′ triiodothyronine, thyroxine). Most of the work is recent, and none older than eleven years, but progress in thyroid research has been so rapid that chromatographic analyses made in 1953 did not in most cases carry either diiodothyronine or triiodothyronine as standards. For this reason slightly older information provides little or no enlightenment with regard to these two important thyroid compounds.

Iodine Metabolism in Nonvertebrates

It appears a safe generalization that of the many organisms that can be shown to metabolize iodine, all make at least monoiodotyrosine and diiodotyrosine with this element. This is even true of marine algae, although here the mechanism of iodide accumulation and oxidation of iodide to iodine may perhaps differ from that in the thyroid gland. However, despite the possible difference in mechanisms, the first iodinated amino acid formed is monoiodotyrosine.

Mono- and diiodotyrosine, as well as iodinated histidines, have been reported as part of iodoproteins throughout the invertebrate world. A high percentage of the protein-bound iodine is in the horny or fibrous structures, but some is in the soft organs—most often in epithelial tissues. Granting the iodination of tyrosines, one might expect thyroxine to be occasionally formed, but it was a surprise to find that both the bivalve *Musculium partumeium* and a number of insects have as much of the total tracer iodine in the form of thyroxine as a mammal. A small part of the high proportion of thyroxine in insects could have been due to unseparated triiodothyronine. The cockroach, the only insect examined organ by organ, concentrated thyroxine mainly in the nerve cord, fat body, and muscle. Nerve cord is of special interest since, in the mammal, thyroxine is both accumulated and metabolized there. Mammalian skeletal muscle, too, has been found to have a protein that binds thyroxine. It is not yet known whether thyroxine is produced in the nerve cord and muscle of the insect, or is produced elsewhere—nerve cord and muscle being sites of concentration. Two other thyronines have been identified chromatographically in invertebrates, 3:5:3′ triiodothyronine in a gorgonian, *Eunicella verrucosa stricta*, and both triiodothyronine and diiodothyronine in the snail, *Planorbis corneus*, an unusual animal since it made two relatively rare iodinated thyronines, but no thyroxine.

These facts, demonstrating the widespread occurrence of thyroid hormones and of hormonal precursors in thyroidless invertebrates, must be taken into account in any theory of evolution of thyroid function. The mere occurrence of iodoproteins, even in high concentration, cannot be considered as evidence that their thyroxine, or thyroxine-like constituents are being used in any particular invertebrate organism. In fact, no clear evidence has yet been presented to show that treatment of an invertebrate with thyroid hormone results in any kind of response. Thus, the most basic properties of thyroid glands, iodide transport, and protein binding of iodine, occur for no known purpose in many plant and animal organisms. The question of the evolution of a special vertebrate organ containing these properties is, therefore, really one of development of a metabolic use for this common substance, thyroxine.

Radioiodine Metabolism in Fish

Only two aspects of thyroidal metabolism of fish will be considered: radioiodine accumulation and radiohormone production. In general, thyroids of salt-water fishes living in an iodine-rich environment accumulate less radioiodine and thyroids of freshwater fish more, but there is a considerable range of overlap. The least efficient freshwater thyroids, i.e., those with the lowest avidity for iodine, were found in the goldfish, *Carassius*, and the sunfish, *Lepomis*. Chavin (1956) has shown the goldfish to have heterotopic thyroid tissue in the head kidney, and when both thyroids were taken into account the goldfish accumulated almost as much of an injected dose of radioiodine as the iodine-starved freshwater fish of the Great Lakes area. A more detailed quantitative study in different seasons of *Fundulus heteroclitus*, a fish that can live in fresh as well as salt water, has resolved some questions. Thyroidal

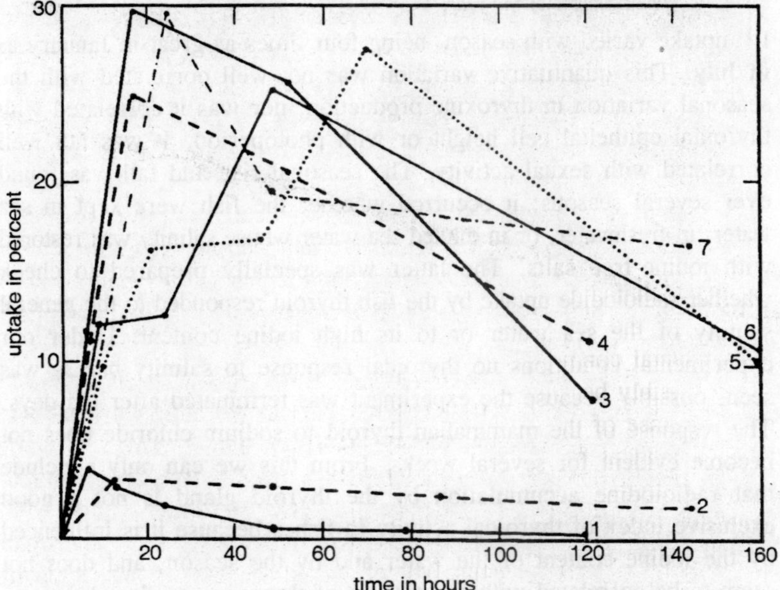

Fig. 5.1. Radioiodine uptake and loss over a period of time in seven species of fresh water teleosts maintained at 21°C. 1. Carassius; 2. Lepomis gibbosus; 3. Percina caprodes; 4. Notropis deliciosus; 5. Umbra limi; 6. Umbra pygmaeus; 7. Xiphophorus maculatus.

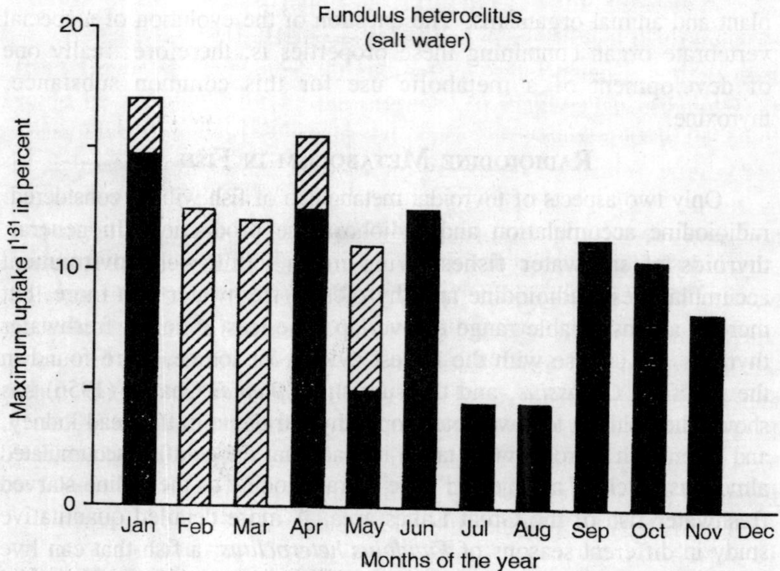

Fig. 5.2. Maximum accumulation of radioiodine by the thyroid of the killifish during eleven months of the year.

I^{131} uptake varied with season, being four times as great in January as in July. This quantitative variation was not well correlated with the seasonal variation in thyroxine production, nor was it correlated with thyroidal epithelial cell height or with photoperiod. It was not well correlated with sexual activity. The seasonal rise and fall was found over several seasons; it occurred whether the fish were kept in sea water, in freshwater, or in diluted sea water whose salinity was restored with iodine-free salts. The latter was specially prepared to check whether radioiodide uptake by the fish thyroid responded to the general salinity of the sea water or to its high iodine content. Under our experimental conditions no thyroidal response to salinity *per se* was seen, possibly because the experiment was terminated after ten days. The response of the mammalian thyroid to sodium chloride does not become evident for several weeks. From this we can only conclude that radioiodine accumulation by the thyroid gland is not a good exclusive index of thyroidal activity in fishes because it is influenced by the iodine content of the water and by the season, and does not seem to be correlated with the amount of thyroxine produced.

Thyroxine has been found in almost every fish tested, from the ammocoetes larva to the lungfish *Protopterus*. Our own negative results with several species may be accounted for, at least in part, by the

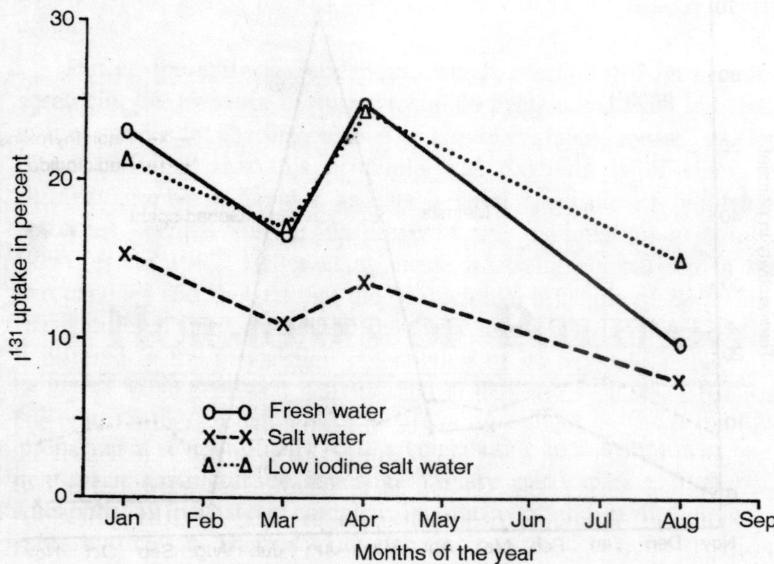

Fig. 5.3. Radioiodine uptake and loss over a period of time in Fundulus heteroclitus.

fact that thyroxine is often slow to appear and is present only in trace amounts or only in certain seasons. Looking for a seasonal pattern in thyroxinogenesis we analyzed iodine utilization in *Fundulus heteroclitus* each month for a year. *Fundulus* made no newly labeled thyroxine within test periods of more than five days in August through November; its thyroxine production started slowly in January, was steady until April, reached a peak in June, and tapered off until it reached zero again in midsummer. The above data were from *Fundulus* kept in salt water, but *Fundulus* simultaneously captured and kept in fresh water in March made about as much radiothyroxine. In this laboratory experiment, temperature was not a factor for thyroxine production, since under our experimental conditions all fish were acclimatized and maintained at the same constant temperature. The *Fundulus* with the most active thyroids (in respect to thyroxine production) were actively breeding, and those with quiescent thyroids had spent gonads, so sexual cycle correlates well with thyroxine production. Photoperiod is another possibility, but was not tested separately.

Triiodothyronine has proven difficult to demonstrate in fish. It was present in *Fundulus* only when the gland was most active, and then only in trace amounts, and has been demonstrated in *Protopterus* in relatively small proportion. We were, therefore, surprised to find one group of mud minnows, *Umbra limi*, which made a considerable

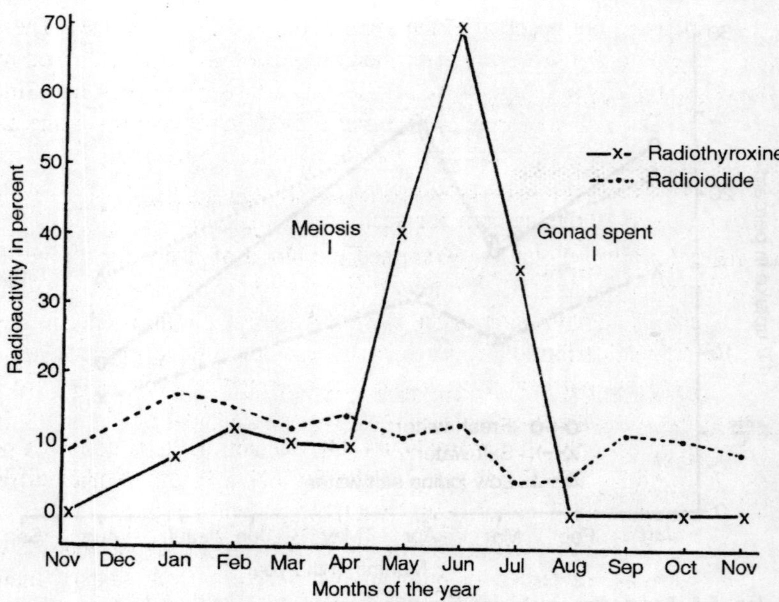

Fig. 5.4. Maximal radioiodine uptake and maximal radiothyroxine production in Fundulus heteroclitus during eleven months of the year.

amount of 3:5:3´ triiodothyronine, and no thyroxine. Three series of experiments were conducted, one in Michigan in August, using Michigan *Umbra limi*, and two in New York in March and December, using *Umbra* shipped from Minnesota and Wisconsin. In the Michigan series, 54 hours after injection of fish kept at 15°C, about 7 per cent of the thyroidal 1^{131} was converted to radiotriiodothyronine, but no radiothyroxine was produced. At room temperature, at approximately the same time interval (48 hours) another series of fish from Minnesota had converted about 10 per cent of the I^{131} to radiothyroxine in addition to 4.5 per cent in the form of radiotriiodothyronine. Two groups of *Umbra* from Wisconsin kept at different temperatures made no appreciable amounts of radiotriiodothyronine, but at the lower temperature labeled thyroxine production was very high. Leloup and Lachiver (1955) suggest that a paucity of environmental iodine may determine triiodothyronine production. This condition certainly prevailed for the mud minnows in Michigan, where iodine was low, while the requirement may have been quite high, as the fish were used shortly after breeding.

Temperature was not a variable in these experiments, but there are indications that it has an effect on hormonogenesis in nature—

although not necessarily the same effect in different species. The *Umbra* mentioned above, which formed radiotriiodothyronine, formed more of it at a lower temperature and less at a higher temperature. Similarly, the December-March *Umbra limi* formed more new thyroxine at lower temperatures and less at higher temperatures. This may be an example of temperature acclimatization in poikilothermic organisms and suggests a possible mechanism for such acclimatization.

Only Leloup has examined the blood of fishes for iodine. He has demonstrated iodide in the red blood corpuscles, and iodide, thyroxine, and triiodothyronine in the plasma. There is no uniformity in the blood iodine distribution in various fishes—some have erythrocytes which are very permeable to iodide, in others the erythrocytes are almost free of iodine. In some species thyroxine appeared rapidly in the plasma, in others only after many hours. Once again, the tests defined a possibly transitory functional state of the thyroid and will remain difficult to interpret until patterns are available for all seasons.

With our present knowledge there is no indication that the thyroid hormone in fishes is qualitatively different from that of mammals. The peripheral target of the hormone is unknown, and so is its physiological effect.

Iodine Metabolism in Amphibia

Frogs are remarkably sensitive to thyroxine analogues—as judged by their metamorphic rate. Triiodothyropropionic acid is the most potent of all the analogues, being 300 times more efficient than thyroxine in *Rana pipiens* and 130 times more effective in *Rana catesbeiana*. In general the 3:5:3′ triiodinated analogues are more effective than their tetraiodinated counterparts, just as triiodothyronine is usually more effective than thyroxine. In addition to the analogues, injections of iodinated proteins speed up metamorphosis.

In species that respond so sensitively to the analogues the question of the nature of the "true" hormone is raised. It is necessary to remember that the exceptional potency of an analogue may be an indication that it is not a natural metabolite. A receptor cell must have both a means of binding a hormone and a mechanism for removing it or its waste product. If the analogue is bound in the same manner as the hormone but is not removed by the usual cellular mechanism, then a small quantity may give continued stimulation creating the illusion of superior specificity. Therefore, no analogue should be considered as a possible thyroid hormone until it can be shown that it is made by the animal.

As part of a larger study on iodine metabolism in neotenous urodeles, we implanted thyroxine pellets into *Amphiuma means* adults. Two weeks later, they were given a tracer dose of 1^{131}, and 2 to 4 days after injection they were sacrificed. On the basis of these experiments, the thyroid of *Amphiuma* is very inactive. Although the salamanders were kept at 20°C, iodine accumulation was very slow, and the steps of conversion to thyroxine even slower. Ninety-six hours after injection there was more mono- than diiodotyrosine, and no triiodothyronine or thyroxine. This may be directly referable to the season (the experiment was done in winter). In any event, thyroxine is not without effect in *Amphiuma*, since it decreased radioiodine uptake quite sharply even below its normal low value of less than 1 per cent.

A similar study of *Necturus maculosus*, a neotenous salamander that retains its gills, showed an even more sluggish thyroid than that of *Amphiuma*. A tracer dose of 20 μc of I^{131} was given one week after implantation of thyroxine pellets into some of the *Necturus*, and the animals were sacrificed 24 to 192 hours later. Uptake of radioiodine was very low (maximum, 1.5 per cent of the injected dose), and no thyroxine at all was produced within eight days after injection. It would be tempting to conclude that neoteny in *Amphiuma* and *Necturus* is due to thyroidal inactivity. However, as is well known, neither of these species responds to treatment with large doses of thyroxine. These tests were made from April to June; the mating season is in the fall. There is some indication, on the basis of radioactive tracer studies, that in a salamander, as in mammals, thyroxine is excreted through the liver. Schmidt (1956) has shown some organic iodine in the bile of *Ambystoma gracile*; the chemical form of the biliary iodine is not yet known.

Radioiodine Metabolism in Reptiles and Birds

The thyroid gland of turtles kept without water, although very slow, eventually accumulates the most iodine of an injected dose of any vertebrate species tested to date. According to Shellabarger et al. (1956) the unusual maximum uptake of 80 per cent of the injected dose of I^{131} may be made possible by reabsorption of radioiodine from the urinary bladder, which is seldom voided when the turtles are kept in a dry environment. There is a correlation between season and thyroidal iodine binding, but the high iodine accumulation is not linked with a high thyroxine production. A complete seasonal study has not been made, but in turtles, as in fish, there are indications that thyroxine is produced in some seasons and not in others.

Both species of turtles tested by Shellabarger et al., one terrestrial the other aquatic, behaved similarly in terms of thyroidal accumulation of radioiodine during periods of dessication. To learn whether this is a feature common to other reptiles, we have completed a series of tracer-iodine studies, under wet and dry conditions, with two lizards: *Anolis carolinensis* and *Sceleporus occidentalis*. Peak uptakes under dry conditions of 12 per cent and 25 per cent, respectively, are well in line with nonreptilian species. The absence of water does not seem to have the effect on lizards that it has on turtles. Both species of lizards produced some thyroxine and a trace of triiodothyronine, differing again from the turtles. In addition, thyroids of both lizards produced traces of two unknown substances. In the lizards both unknowns were present in minute quantities; they were of possible interest because they were common to both species. Furthermore, these substances characterized

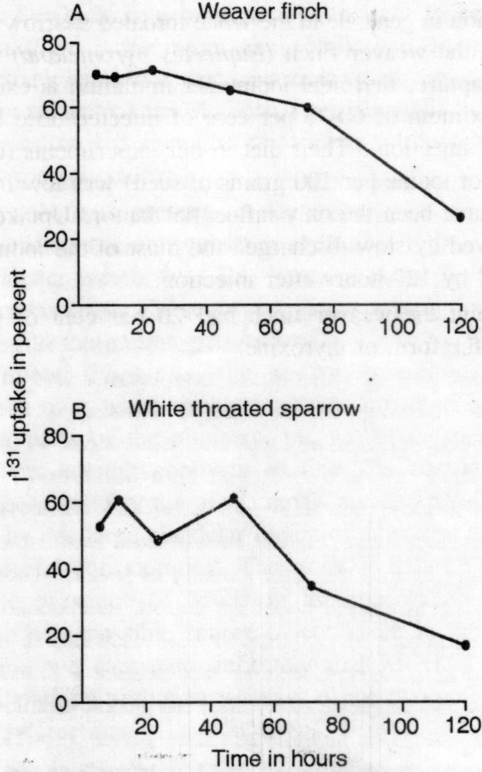

Fig. 5.5. Radioiodine uptake and loss over a period of time in the weaver finch and the white throated sparrow.

clearly by chromatographic analysis, were found not only in lizards, but also in even greater proportions in two species of birds tested in our laboratory.

There have been several reports on utilization of radioiodine by chicks, the most extensive being a study of cockerels by Vlijm (1958). Iodine utilization by the thyroid of the adult cockerel is not remarkable, except that the cockerels produced 3:5 diiodothyronine as well as mono- and diiodotyrosine, 3:5:3´ triiodothyronine, and thyroxine. Both the triiodothyronine and the diiodothyronine deserve comment. The former has also been identified recently in the chick by Shellabarger and Pitt-Rivers (1958). Its presence is not unexpected, but: the fowl is the only form tested in which the triiodinated compound is not more potent than its tetraiodinated homologue, thyroxine.

The domesticated chick may not be a very good index of avian thyroidal function in general. In the white throated sparrow (*Zonotrichia albicollis*) and the weaver finch (*Euplectes pyromelana*) tested by us shortly after capture, thyroidal iodine accumulation is extremely high and rapid (maximum of 60-70 per cent of injected dose less than six hours after I^{131} injection). Their diet in our experiments (canary seed, 2 micrograms of iodine per 100 grams of seed) was low in iodine, but this may not have been the only influential factor. Uptake was rapid, and was followed by slow discharge, and most of the iodine was gone from the gland by 120 hours after injection.

At one point the weaver finch had 20 per cent of its thyroidal radioiodine in the form of thyroxine.

6

Renal and Heterotopic Thyroid Tissue

Thyroid tissue occurring in areas of the body remote from the usual pharyngeal location has been described in a number of vertebrates. Such occurrences have usually been attributed to metastasis from pharyngeal thyroid tumors. In the case of "struma ovarii," thyroid tissue in the human ovary, the origin of the thyroid tissue has been regarded as teratomatous.

Recently, Baker et al., (1955) and Chavin (1956a, b, c) have reported occurrence of hyperplastic or normal thyroid tissue in the kidneys of aquarium-reared platyfish, *Xiphophorus maculatus*, and goldfish, *Carassius auratus*. In both instances renal thyroid tissue was found in a high proportion of the fish studied. Further studies on platyfish led to the conclusion that thyroid tissue in this species was capable of an extraordinary degree of mobility, leading to its dispersion from the pharynx to numerous distant areas of the body. It is thought that the migratory ability shown by fish thyroid may have some bearing on thyroidal behaviour in other vertebrates, particularly man.

Development of Heterotopic Thyroid Tissue in Platyfish

Several strains of platyfish were found to be subject to the development of thyroid tumors in the kidneys, concurrently with the appearance of pharyngeal thyroid hyperplasia. One of these strains (*BH*) was selected for intensive investigation of the tumors' genesis. Radioiodine autography of serial sections was used to locate thyroid tissue throughout the bodies of these fish, at various stages in their development.

Heterotopic thyroid tissue was not found at, or before, birth (platyfish are viviparous) and did not appear before the fish were about two months old. After that age, the number of fish with renal thyroid follicles increased with time. These follicles were morphologically normal. Their number per fish also increased with age; they were present in all parts of the kidneys, but the highest concentration was found near the cardinal veins, which return blood from the renal portal system to the heart. As they grew still older, fish which had renal thyroid follicles also developed normal thyroid tissue in several other heterotopic sites. The most frequent of these were the spleen, the chorioid gland in the eye, and the ventricle of the heart. Infrequently occupied sites included the air bladder gland, brain, base of gills, connective tissue near the ear and pseudobranch, and intestinal wall. As the fish aged, thyroid follicles in the pharynx also became more numerous and occurred throughout a larger area. This is typical of

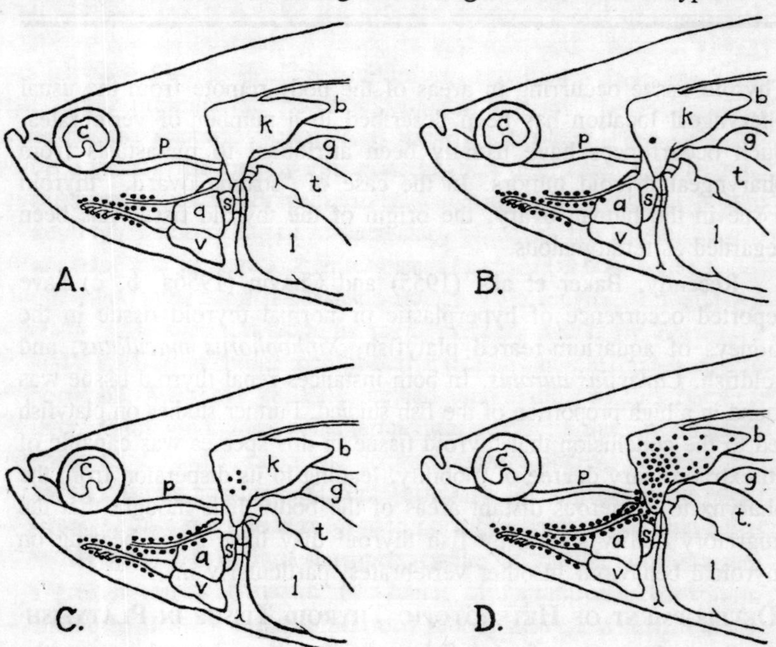

Fig. 6.1. The increase in the extent of thyroid distribution, as age increases, in strain BH platyfish. A—One month old, average of four fish; B—Three months old, average of six fish; C—Six months old, average of seven fish; D—Ten months old, average of four fish. c, chorioid gland; p, pharynx; a, auricle; s, sinus venosus; v, ventricle; k, kidney; b, air bladder; g, gall bladder; t, stomach; l, liver.

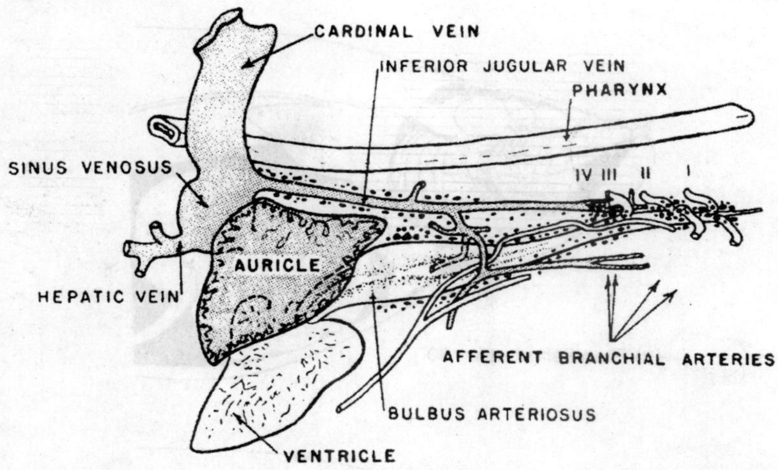

Fig. 6.2. The distribution of thyroid follicles in relation to blood vessels and associated structures in the normal adult platyfish.

growing teleosts in which most species have a nonencapsulated thyroid "gland." In the platyfish, however, the pharyngeal thyroid follicles were found to spread further than had been considered normal. From their original embryonic position along the ventral aorta, between the third and fourth branchial arches, thyroid follicles spread laterally towards the gill bases and caudally into the pericardium around the bulbus arteriosus and above the heart. They appeared to follow the inferior jugular vein caudally, clustering around it, and usually were present in large numbers at the point where this vein enters the cardinal vein. In many normal adult fish, thyroid follicles appeared on the surfaces of the cardinal veins near their point of incorporation into the kidneys. Thyroid tissue also was found on the surface of the bulbus arteriosus very near the heart and on the pericardial surface.

When the *BH* strain platyfish reached the age of about 9 months, thyroid tissue in some of them began to show hypertrophy. In time, progressive hypertrophy led to development of large cystic tumors in the kidneys, accompanied by solid hyperplasia of the pharyngeal thyroid, large thyroid masses in the spleen, and lesser amounts of thyroid in the eye and heart. The tumorous thyroid material of the kidneys and of the pharynx usually became amalgamated around the walls of the cardinal veins, in consequence of the pretumorous distribution of normal thyroid tissue. The average age at which these tumors developed in the *BH* strain of platyfish was about 21 months; the maximum life span of the fish is about 3 years.

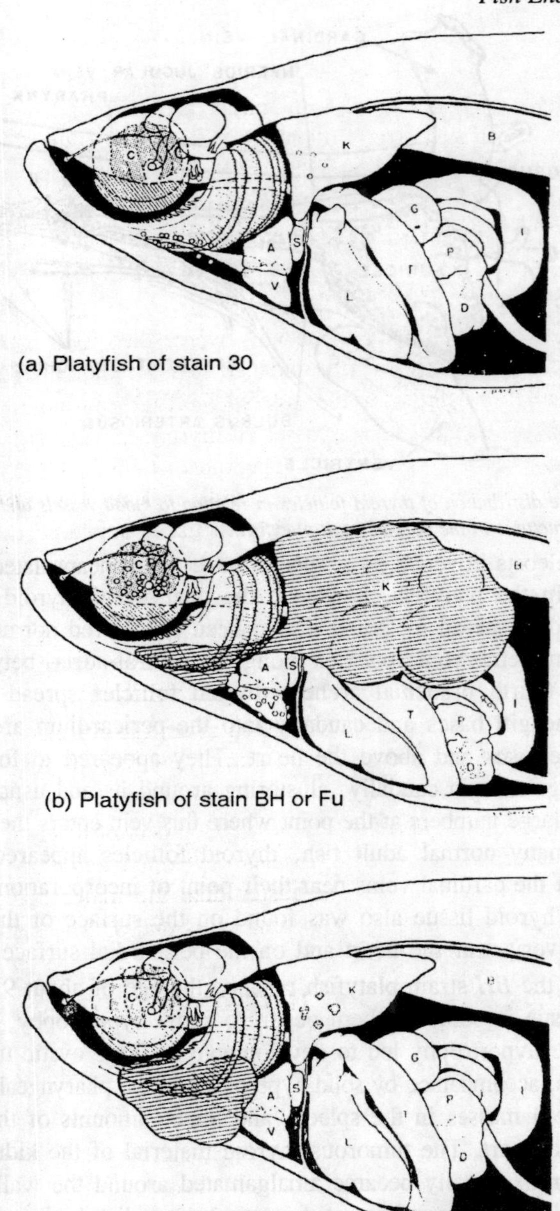

Fig. 6.3. Stain and species differences in thyroidal development in xiphophorin fishes. c, chorioid gland; a, auricle; v, ventricle; s, sinus venosus; k, kidney; l, liver; t, stomach; g, gall bladder; p, pancreas; d, spleen; i, intestine; b, air bladder.

The thyroid tumors in the kidneys were always accompanied by the presence of large cysts, with solidly packed tumor cells occurring in clumps of varying size. Colloid formation or retention was variable: radioiodine autographs showed that the material in the cysts contained little or no protein-bound iodine, and the I^{131} present was generally inversely proportional to cyst size. Heaviest radioiodine binding occurred where massed hypertrophic thyroid cells formed follicles which held some dense eosinophilic colloid.

The spleens of these fish usually were filled with solidly packed hypertrophied follicles, which produced strong radioiodine autographs. Occasionally, cysticity in the spleen was very pronounced. In the chorioidea of the eye, large masses of thyroid tissue were rare; when found they caused a "pop-eyed" appearance. Ordinarily, thyroid tissue in the chorioidea was present in moderate amount. Only one case was found in which a large mass of thyroid tissue appeared in the heart; this mass was not embedded in the heart tissue, but was on the surface of the ventricle.

The pharyngeal thyroid tumors in these fish were rarely very cystic, in contrast to those in the kidneys, and were never large enough to be visible externally or to interfere with gill structures.

In addition to heterotopic sites noted in normal *BH* platyfish, thyroid tissue was occasionally found in tumorous fish in the braincase, pseudobranch, and liver.

A sex difference appeared in the development of heterotopic thyroid tissue in strain *BH* platyfish: females developed more of the normal renal thyroid tissue earlier and faster than the males and ultimately had twice as many renal thyroid tumors as males. The source of this difference was not determined.

Physiological Studies of Heterotopic Thyroid Tissue in Fish

Goldfish

Chavin's studies (1956a, b, c) on thyroid follicles found in the head kidneys of 85-90 per cent of hatchery-raised goldfish showed that the renal thyroid tissue had a much slower uptake of inorganic radioiodine than did the pharyngeal thyroid tissue of the same animals: it reached a maximum after 24 hours, in contrast to the one hour required by the pharyngeal tissue. Both areas retained their maximal uptake over a 20-day period after injection of the radioiodine. The thyroid in the kidney, on the average, accumulated more of the initial dose than did that in the throat. In 47 per cent of normal goldfish the renal follicles accumulated more radioiodine than those in the pharynx, in 23 per cent of the fish the reverse was true, and in 20 per cent the areas were equal in uptake. In 10 per cent of the goldfish there was no renal thyroid tissue, and consequently no radioiodine accumulated in these organs. No correlation was found between renal thyroid follicle number and radioiodine uptake in the kidney.

When radioiodine was given as radioiodinated serum albumin, the pharyngeal thyroid reached a maximum uptake in 24 hours, but the renal thyroid follicles continued to take up more iodine for 20 days after the injection, again reaching a slightly higher figure than the throat thyroid.

The renal thyroid follicles in the goldfish reacted similarly, in their radioiodine uptake and cellular morphology, to those in the throat when the fish were hypophysectomized or treated with thyroxine, cortisone, thyrotropin, thyrotropin plus thyroxine, thiouracil, or saline.

Platyfish

No studies comparing the activities of pharyngeal and heterotopic thyroid have been completed in the platyfish. Radioautographs of normal fish showed strong radioiodine binding by follicles in both areas after 24 hours of immersion in tracer radioiodine solutions.

In the platyfish, however, the very existence of heterotopic thyroid follicles was found to depend upon the amount of iodine available to the developing fish. Throughout the laboratory in which these fish were bred, the aquarium water contained one-third, or less, the iodine concentration of New York City tap water. It was under these conditions that normal and goitrous heterotopic thyroid tissue developed in the platyfish and, also, that pharyngeal thyroid tumors developed in Montezuma swordtails.

When strain *BH* platyfish were kept in aquarium water to which 1 mg/l of potassium iodide had been added, the development of normal renal thyroid tissue was halted at the stage it had reached when treatment was begun. Thus, platyfish raised in the iodide-enriched water starting from any age under two months were free of heterotopic thyroid tissue as adults, although untreated siblings had large amounts of this tissue. When treatment was begun at more advanced ages, heterotopic thyroid tissue in the treated fish was comparable to that in siblings sacrificed at the beginning of treatment; untreated siblings were found to have much more renal and other nonpharyngeal thyroid tissue and the beginnings of renal thyroid tumors.

It, therefore, seems evident that heterotopic thyroid in platyfish, like pharyngeal goiter in many vertebrates, develops as a compensatory device when the iodine demands of the animal exceed the available supply of this essential component of thyroid hormone. It is notable that this response of thyroid tissue preceded, by a long time interval, any hypertrophy of the thyroid cells, the commonly found response of thyroid tissue to insufficient iodine.

Iodide treatment of platyfish which had large renal thyroid tumors led to involution of the thyroid tissue, as it did in pharyngeal thyroid tumors of Montezuma swordtails. Thyroid tissue did not disappear entirely from the kidneys with prolonged treatment, but remained in the form of many small inactive follicles scattered throughout the kidney tissues. Extensive destruction of kidney tissues during the development of the tumors prevented these organs from making a complete recovery.

Prolonged thiourea treatment of young adult *BH* platyfish, beginning at an age when a number of normal thyroid follicles should have been present in the kidneys, elicited the premature development of goitrous thyroid tissue in these organs; cystic masses in some animals paralleled the early stages of the spontaneous renal thyroid tumors. Thyroid tissue in the spleen, heart, and chorioid gland of the thiourea-treated fish became hypertrophied in a way similar to their pharyngeal thyroid tissue. There was some evidence that the thiourea treatment led to an increased heterotopic distribution of thyroid tissue.

Most of the data on physiological responses of heterotopic thyroid tissue in goldfish and platyfish tend to show that this tissue is not basically different from the pharyngeal thyroid tissue of the same animals. The main differences were that (1) in the goldfish, renal thyroid had a slower radioiodine uptake than pharyngeal thyroid, and (2) in the platyfish, thyroid tumors in the kidneys tended to become more cystic than those in the pharynx. The first of these differences might be caused by the removal of much radioiodine from the blood before it reached the kidneys. Intraperitoneally absorbed radioiodine, to a large extent, may have passed through the hepatic portal system, thence to the heart, gills, thyroid, and general circulation before reaching the renal portal system; direct absorption by the kidneys of intraperitoneal radioiodine might be slight, since these organs are outside the peritoneum. The pharyngeal thyroid may thus have become saturated with radioiodine immediately, while the kidney thyroid continued to take up radioiodine, coming from various parts of the body, for a longer period. The second difference was not true of all heterotopic thyroid tissue; for example, splenic thyroid tumors rarely became cystic in fish with cystic renal tumors. Differences between the extent of hyperplasia and hypertrophy of thyroid tissue in different organs of the same individual seem more likely to result from differences in the local environment than from intrinsic factors. For example, Baker (19580) depicted hypertrophied and quiescent thyroid follicles in the eye of the same goitrous platyfish, but the hypertrophied follicles were embedded in the chorioid gland, where they may have received more contact with the blood than the quiescent follicles attached to the medial surface of the retina. Assuming this to be so, the follicles in the chorioid gland thus have received more thyroid-stimulating hormone than those on the retina.

Origin of Heterotopic Thyroid Tissue in Platyfish

Four ways in which heterotopic thyroid tissue might have arisen in platyfish were postulated by Baker (1958a). These were: (1) by displacement of portions of the thyroid anlage during embryogenesis; (2) by the differentiation of embryonic "rest" cells remaining in the adult; (3) by vascular transport of thyroid cells or fragments from the pharyngeal area; or (4) by active migration of thyroid cells or follicles from the pharyngeal area. The first of these possibilities seemed to be eliminated when it was found that platyfish did not develop heterotopic thyroid tissue until at least two months after birth, although the pharyngeal thyroid is fully differentiated and functional prenatally.

The second possibility was made less likely, and the third and fourth more probable, by the results of experiments in which the pharyngeal thyroid of very young platyfish was destroyed by exposure to large amounts of radioiodine. These fish were too young when treated to have had any heterotopic thyroid tissue. It was reasoned that if heterotopic thyroid follicles arose from undifferentiated and unrecognized cells which bound no radioiodine, removal of the pharyngeal thyroid should not interfere with (but, if anything, promote) their differentiation into thyroid tissue. On the other hand, if heterotopic thyroid arose by metastasis or migration from the throat, no such tissue should appear later in the thyroidectomized fish. No heterotopic thyroid tissue did appear in fish so treated, although it was found in a high proportion of untreated siblings.

The third possibility, that of a metastatic origin of heterotopic thyroid, was never excluded, although no follicular emboli or thyroid-like cells were ever seen in blood vessels. It seemed unlikely that pharyngeal thyroid cells in normal fish should have an impetus to seek entry into the blood vessels in that area, as the follicles were not crowded in the younger fish. As the blood vessels in the area are mostly thick-walled arteries, such entry might have been difficult. Metastasis to the sites where heterotopic thyroid tissue was found most often was made less likely by the direction of blood flow in the vessels entering or draining the various areas. For example, venous blood from the pharyngeal thyroid area passes through the inferior jugular vein, enters one of the cardinal veins below the kidney, and flows to the heart; thyroid emboli entering here should pass into the heart and thence directly to the gills and pseudobranch, and should not tend to pass into the kidney. Yet in normal fish very little thyroid tissue was found in the heart and none in the gills or pseudobranch, but large amounts were found in the kidneys. Also, the kidney whose cardinal vein received the inferior jugular vein (this is variable) was not found to contain significantly more thyroid follicles than the other kidney.

The fourth possibility, that of the origin of heterotopic thyroid tissue by active migration, remained for consideration. This means of dispersion was rendered likely by the observed increased spread of pharyngeal thyroid follicles as the fish grew older, and, particularly, by the tendency of the follicles to appear along the surfaces of the blood vessels near the kidneys. It was simple to postulate that follicles or cells occasionally might migrate further than points reached by the main pharyngeal mass, using blood vessels or connective tissue as a substratum, and enter the kidneys near the points where the cardinal veins left these organs. Follicles were found clinging to the surfaces of the kidneys and cardinals at these points.

The vascular arrangement of the platyfish could provide direct pathways for dispersal of thyroid tissue to the kidneys and heart from the pharyngeal area, and from the kidneys to the chorioid gland, via the jugular and ophthalmic veins, if thyroid cells or follicles migrated actively. The use of blood vessels as a substratum for either passive or active migration has been noted by Gudernatsch (1911) and others. Owing to the absence of thyroid follicles along most portions of the postulated pathways to various organs, it seemed more likely that migration, if it occurred, was a property of single thyroid cells, not whole follicles, and that colloid secretion and follicle development usually took place only when favourable environments were reached.

It was considered that blood transport may have carried thyroid fragments from the kidneys into the heart, once thyroid follicles were established in the former. It will be recalled that in normal platyfish no thyroid tissue appeared in the heart, spleen, or chorioid gland until it had first appeared in the kidneys.

Stain and Species Differences in Heterotopic Thyroid Development and Genetic Studies on Renal Thyroid Tumor Development in Platyfish

Platyfish

Thyroid tumors in the kidneys were found in individuals of seven, more or less inbred, strains of platyfish. At least three of these strains had a relatively high incidence of these tumors under laboratory conditions. One highly inbred strain (30) was shown to develop no thyroid tumors in the kidneys or pharynx, although it was derived from the same ancestors as a strain (163) which did develop renal thyroid tumors. None of the strains of platyfish examined were entirely free of individuals which had normal thyroid tissue in their kidneys.

The *BH* strain had a renal thyroid tumor incidence very close to that of the related *Fu* strain, yet the sex distribution of tumor-bearing fish in these two strains was completely reciprocal. In the *BH* strain twice as many females as males developed the tumors, but in the *Fu* strain exactly the opposite was true.

Analyses of F_1 and F_2 hybrids from crosses made between the 30 and *BH* strains showed that tumor development was independent of the dominant sex-linked factors present in the cross. The F_1 had a higher incidence of the tumors than the *BH* parent stock, which showed that genes related to tumor development were not recessive, and possibly indicated that an interaction of genes had enhanced tumor development. The F_2 had a considerably lower tumor incidence than the *BH* stock. The tumor incidences in the two generations considered together suggested the action of multiple dominant factors in tumorigenesis.

Other Species

Laboratory-bred Montezuma swordtails were highly susceptible to large pharyngeal thyroid tumors. Examination of fish of this species for renal thyroid tissue showed that although this tissue occurred, it did not develop to the extent found in some of the platyfish strains. Only one thyroid tumor in the kidneys was found among Montezuma swordtails, and only one-third of those with pharyngeal thyroid tumors had thyroid in their kidneys.

Thus three distinct patterns of response to the same low iodine environment were found in xiphophorin fishes: (1) no thyroid tumors developed in any part of the body (strain 30 platyfish); (2) large pharyngeal thyroid tumors developed, without heterotopic thyroid tumors (Montezuma swordtails); and (3) heterotopic thyroid tumors developed very extensively, but pharyngeal thyroid tumors were not prominent (strain *BH* and *Fu* platyfish).

Individual fish of three other species with heterotopic thyroid tissue were found. The guppy was different from platyfish in that it had many normal thyroid follicles in the spleen but no thyroid tissue in the kidneys. The cherry barb and spike-tail platyfish had thyroid tumors in both pharynx and kidneys, as in strain *BH* platyfish.

7

ULTIMOBRANCHIAL GLAND

Among derivatives of the embryonic pharyngeal entoderm are certain small, usually paired glandular structures best known today as the ultimobranchial bodies. These "bodies" were originally regarded as "lateral" or accessory thyroids because in typical mammals they were found to join with the lateral lobes of the thyroid gland. They were also known as suprapericardial bodies, because of their position in elasmobranchs and in certain amphibians, and as postbranchial bodies or as telobranchial bodies.

The term "ultimobranchial body" is to be preferred, since it may be used in any class of vertebrates and implies simply an origin similar to that of the pharyngeal pouches but does not imply that this outpocketing represents any particular pouch. The varied development, differentiation, and fate of this structure make it of peculiar significance and render it subject to different interpretations. No functions have been proven, although several have been hypothesized.

Occurrence

The comparative embryology of the ultimobranchial bodies has been extensively investigated. They are present among all the orders of each class of vertebrates except the cyclostomes. With some exceptions in teleosts, ultimobranchial bodies have been noted in fishes. According to Giacomini (1908) and Watzka (1933), the ultimobranchial bodies can be considered homologous in all teleosts and with similar structures in selachians also, but they are difficult to render homologous throughout the various, entire series of vertebrates.

In amphibia, these structures are almost universally present among anura, and Wilder (1929) found them without exception in urodeles.

They are present in virtually all reptiles, birds, and mammals, being reported absent only in some snakes and in the bat and horse. They are always paired in mammals but may be unpaired in other vertebrates. When unpaired, it is usually the left that is present (urodeles; some reptiles and birds). In most nonmammalian forms, the fourth and fifth pouches and the ultimobranchial anlagen develop as united blind outpocketings on each side of the posterior part of the pharynx. In typical mammals, however, it appears uniformly on both sides behind the fourth pouch, except in the rat and mouse where the third pouch is the last.

If these structures are truly ultimobranchial, they cannot be considered directly homologous throughout the different classes of vertebrates, where the number of arches and pouches in the series may vary. However, there seems to be no reason to force a homology too far. Whatever the factors involved may be, they affect the caudal branchial region and in most vertebrates a structure develops that is comparable if not homologous.

Character in Nonmammalian Forms

Ultimobranchial tissue has a vesicular structure in fishes. In elasmobranchs (*Squalus acanthias*), Camp (1917) found that this so-called gland consists of large, distended vesicles which frequently intercommunicate and contain mucous secretion. In teleosts, Giacomini (1908) found granular substance mixed with degenerated nuclei, and in both groups Watzka (1933) reported a thyroid-like colloid. According to Maurer (1888), a single large follicle or a complex of smaller ones represents this body in anura. During certain developmental stages, this sometimes contains a serous product but never colloid. In urodeles, although variable in size, form and position, it occasionally exhibits considerable secretory activity of variable quality and may bear a superficial resemblance to thyroid tissue. It remains throughout life as an epithelioid or epithelial structure which is frequently vesicular. The differentiation of its cells, however, is frequently so slight that there is the suggestion of persistence of an essentially embryonic structure. In urodeles, there is little or no evidence to support the assumption that the structure in question possesses physiological significance, either as an internally secreting gland or as an exocrine gland that has lost its duct. Its inactivity or inhibition during development suggests no correlation with metamorphosis in amphibia. However, contrary to Maurer's statements (1899, 1902) that colloid is absent in all nonmammalian vertebrates, appearing first in *Echidna*, Watzka (1933) noted colloidal substance stores in vesicles of the turtle, lizard and snake.

The ultimobranchial body in reptiles may have an endocrine function, at least in the lizard (*Lacerta*) where it is well developed. During the early postembryonal period, this structure, at first a compact mass of gland-like epithelial cords, becomes well vascularized. With growth and differentiation, there is evidence of secretory activity in numerous small thyroid-like vesicles. These enlarge and produce a fine homogeneous colloid in adult animals. According to Eggert, the activity of ultimobranchial tissue in the lizard seems to parallel seasonal changes. This is especially noticeable in June and July, when secretion begins and there is evidence of mitotic activity. This activity decreases in the fall, and during hibernation the colloid becomes denser and the structure is quiescent. Spontaneous involution occurs with age in the second and third summers and is associated with large secretion-filled cysts. Connective tissue and lipoid material usually replace these follicles and there is infiltration of lymphocytes.

Eggert has subjected this tissue to experimental analysis. No change occurs in animals in which ultimobranchial tissue has been removed. Neither extirpation of the thyroid nor involution of the testes cause demonstrable change in ultimobranchial tissue. Presumably, thyrotropic hormone acts directly on this tissue because prolonged injection of TSH causes increased activity with tremendous secretion and cyst formations in the absence of thyroid tissue. In this respect, ultimobranchial tissue may resemble thymus tissue. Furthermore, observations dealing with metamorphosis following extirpation of the thyroid gland suggests no thyroid-like activity in this ultimobranchial tissue, but simultaneous injection of TSH with thyroxin decreases cyst formation (normally, thyroxin produces no essential change). This author's concept of the nature of ultimobranchial tissue in reptiles may be open to question and has been ignored for years.

In birds the disposition of the last three pharyngeal pouches (IV, V and VI) closely resembles that in reptiles. In these animals the rudimentary fifth pouch is a transitory structure and the sixth pouch (found in sequence) gives rise to the ultimobranchial body. It is also quite likely that these pouches become incorporated or combined as a "caudal pharyngeal complex."

Although Lillie (1919) called attention to the development of thymus and parathyroid tissue from the ultimobranchial bodies in the chick, Dudley (1942) has shown that parathyroid tissue and thymus-like areas occur in addition to typical ultimobranchial parenchyma. The formation of one parathyroid in each ultimobranchial body would agree with a branchiomeric interpretation of the pharynx (which attributes to a posterior rudimentary pouch potentialities of one or two well developed pouches ahead of it—typically pouch III and IV), but the number of glands found is variable, and they may even be multiple. If not a matter of embryonic induction—the presence of parathyroid tissues being linked with the proximity of a parathyroid III and IV—this may mean that entoderm of the last diverticulum has ultimobranchial and parathyroid potentialities. In the presence of numerous lymphocytes, a "lympho-epithelial reaction" similar to thymus tissue may also occur. Verdun noted the "distinctly glandular character of this parenchyma" as long ago as 1898.

Based on the premise that all organs are of use or have been functional in ancestral forms, the evidence presented by Eggert (1938) for an endocrine function in reptiles is the most impressive. In birds, Terni (1927) and Watzka, working independently, both agreed that the ultimobranchial body had all the "morphological requisites of an actively functioning endocrine gland and Terni even believed it to be cyclic or seasonal. However, after an extensive review of the problem,, Dudley (1942) has more recently found no reason to regard this tissue as having an endocrine function in birds. She stated that only rarely, in the fowl, are these structures reminiscent of secretory follicles, thyroid or otherwise. This tissue may possess certain secretory attributes, but as Kingsbury (1915) pointed out, "colloid may signify no more than a retained secretion of high protein content . . . and vesicles or cysts may be expected to arise in epithelial masses originating from a surface epithelium." If it is not a gland, with specific functions, there remains the interpretation offered by Kingsbury: "continued growth activity in the branchial entoderm."

In most mammals, the ultimobranchial body is so intimately associated with the fourth pouch that the two may be appropriately called the fourth or caudal pharyngeal complex. Although in birds and lower vertebrates generally no real transformation into thyroid tissue occurs, the developmental relations of this "complex" and the thyroid become very intimate in mammals, and there is even evidence that this tissue differentiates into colloid-filled vesicles. Echidna is the only mammal thus far reported whose ultimobranchial body never becomes embedded into the thyroid, but differentiates into glandular tissue, which, while differing from the thyroid in structure, nevertheless possesses colloid-containing follicles. Its similarity with that of birds has been noted.

There are many arguments as to the fate of the ultimobranchial body, especially in mammals (including man). After it becomes intimately incorporated within each lateral lobe of the thyroid gland, its further history varies in different species, as has been reviewed. The question naturally arises as to whether or not ultimobranchial bodies take part in the formation and functions of the thyroid gland.

MAMMALS

Concepts

Since the fate of the ultimobranchial body may be variable not only in different species but also among individuals, interpretations of its significance have been both numerous and diverse.

The *alleged fates* of this structure in mammals are as follows: (1) That it differentiates as an inherent "*lateral thyroid*" component and contributes to the general thyroid parenchyma. This was the early concept.

(2) That it is an "*organ of unknown function.*" Verdun (1898) described cysts and cell strands which persisted as "special rests" (rabbit, rat, dog, and mole) and after further study thought the ultimobranchial body might represent a gland of unknown function. In the bird he regarded it as a vestigial gland. Grosser (1910) working with human material stated that "the ultimobranchial body clearly represents a ductless gland which has become rudimentary" but suggested (1912) that it might be a *thyroid growth center* which ultimately disappears. After examining a variety of animals, Watzka (1933) considered the ultimobranchial body as representing in phylogeny an organ of internal secretion of apparently increasing significance up to and including birds. De Winiwarter (1933) held the opinion that in the guinea pig it could form a so-called parathyroid V and thymus V as well as an "assemblage of canals and cavities"; in the cat (1935) he considered it "an enigmatic organ of unknown function." In the developing human thyroid gland, Politzer and Hann (1935) could not decide whether gland-like remnants of the ultimobranchial body indicated developing thyroid parenchyma or "functionally different gland tissue" within the thyroid. They did not believe the ultimobranchial body degenerated. Godwin (1937) thought that in the dog, at least, ultimobranchial tissue might become scattered to form the so-called "parafollicular cells" ("interfollicular" cells or "macrothyrocytes"),

and in 1940 he suggested that "the idea that it may be an 'organ' in the process of appearing is a possibility to be kept in mind."

(3) That it represents "*more or less indifferent tissue*." In the pig and guinea pig, Simon (1896) thought the ultimobranchial body degenerated and disappeared. Rabl (1907) was of the opinion that it degenerated completely in the mole, but in the guinea pig (1922) it persisted without transformation and with little change in postnatal thyroids. Recent work by Klapper (1946) indicates that it finally undergoes degeneration and completely disappears. In a few instances it persists as irregular cystic masses which on occasion may be converted to a thymus IV. Kingsbury reported that this structure usually undergoes degeneration in mammals, disappearing entirely or resulting in the formation of epithelial cysts; in man this occurs "typically without trace." In the calf (1935) he could find no evidence for the origin of true thyroid parenchyma from ultimobranchial tissue, but he stated that in some instances vesicles indistinguishable from thyroid follicles might develop.

Kingsbury has interpreted the ultimobranchial body only as "the expression of a continued growth activity in the pharyngeal entoderm apparently associated with the gill-forming potentialities of the region." He has created a premise for considering the ultimobranchial body as relatively indifferent tissue which may be influenced in its development by variable factors of environment. Upon the basis of this assumption Rogers (1927), studying the rat, considered the partially differentiated ultimobranchial tissue as a natural transplant which, composed of indifferent material, becomes influenced in its further development by the thyroid parenchyma, "differentiating not according to its origin, but in harmony with its surroundings." This induction theory appears to explain the transformation of ultimobranchial tissue into thyroid-like follicles observed in many mammals. According to Badertscher (pig), Rogers (rat), and Godwin (dog), the amount of thyroid tissue derived from the ultimobranchial body is relatively slight.

Relation to the Thyroid Gland

Since the degree of incorporation and intimacy of fusion of the ultimobranchial body with thyroid parenchyma sometimes varies depending upon the growth mechanics involved during development, it may be intimately and extensively associated with thyroid parenchyma, remain in a "vasculostromal" hilus or lie wholly outside the thyroid gland. According to Watzka (1933), considerable structural variation accompanies the development of ultimobranchial tissue, depending upon its position relative to thyroid parenchyma. In the rat, pig, and dog, the ultimobranchial body becomes so intimately associated with the developing thyroid parenchyma that it undergoes reticulation and transforms into vesicles that are indistinguishable from thyroid follicles. In many mammals, however, a variable portion of the ultimobranchial body remains outside the thyroid; there it "shows a different and not so uncharacteristic and apparently meaningless cystic structure as one occasionally finds within the thyroid gland of many mammals". Its architecture sometimes resembles that of mucous acini. Watzka further states that these structures exhibit well the innate growth of potentialities of the ultimobranchial bodies when they are not restricted by incorporation within the thyroid gland. The writer, from his observations on postnatal thyroids, can readily subscribe to such a belief.

Observations and Experimental Studies

My own work with young sheep indicates that the ultimobranchial bodies are usually represented by large, multiple cysts lined by stratified squamous epithelium in the thyroid gland. They have been regarded as arising congenitally due to the failure of the ultimobranchial body (in sheep and cattle) to fuse intimately with the thyroid parenchyma during embryonic development. This cystic manifestation of metaplasia may reflect a deficiency of vitamin A, for sheep and cattle are grazing or pasture animals. The epithelium of these cysts frequently produces cords or clumps of glandlike cells which invade the neighbouring thyroid parenchyma in a manner that often suggests the growth of glandular neoplasms. Depending presumably upon the state of activity of the thyroid gland, these cysts can be the source of two peculiar outgrowths. These may resemble either typical thyroid parenchyma or adenoma-like nodules which may be solid or cystic, and they originate from the basal (germinative) and/or the suprabasal (*degenerating*) cells in this stratified squamous epithelium. The evidence here was interpreted as indicating that this cyst epithelium, indicative of ultimobranchial tissue which lacked inherent thyroid-forming potentialities, is nevertheless capable of producing undifferentiated cells. These can become induced by adjacent "active" thyroid parenchyma (and possibly other factors) to transform into thyroid-like vesicles; some of which ultimately may develop true neoplasms.

Despite the uncertainty as to whether or not the ultimobranchial body differentiates into typical thyroid parenchyma, the problem is more than an academic one. Dunhill (1931), Ward (1940), Cohn and

Stewart (1940), and others have suggested that aberrant thyroid tissue, possibly a derivative of the ultimobranchial body, is the site of origin for so-called *lateral aberrant thyroid tumors*.

According to Dunhill, as a result of imperfect growth and differentiation, misplaced bits of this tissue whether in the lateral neck or within the thyroid gland itself "may foredoom the host to carcinoma from before the day of birth." In this connection it is of interest to note that in the baboon, ectopic ultimobranchial tissue (associated with a thymus IV) apparently can give rise to accessory thyroid tissue. This spontaneous aberrant thyroid tissue originates from the wall of cysts lined by stratified squamous epithelium. It is a manifestation of ultimobranchial tissue which is not incorporated within the thyroid gland and, therefore, not induced by the proximity of thyroid parenchyma to transform into thyroid tissue. As far as the author is aware, this is the only example of accessory thyroid tissue from a portion of the caudal pharyngeal complex left outside the thyroid gland in mammals, and this strengthens the concept that ultimobranchial tissue is a "lateral thyroid." However, since the main thyroid mass showed marked hyperplasia suggesting overactivity, the stimulus for the appearance and growth of this tissue may be due to an increased demand for thyroid function during a previous stress period (adolescence?). As Ward (1940) has suggested regarding the growth of lateral aberrant thyroid tumors, such differentiation of ultimobranchial tissue may well lie dormant until the proper stimulus is furnished (e.g., perhaps increased endocrine demand during pregnancy).

On the other hand, that ultimobranchial tissue represents "more or less" indifferent tissue, which may, on occasion, be stimulated to produce thyroid tissue even outside the thyroid gland has also received support. Thymic adenomata, resembling accessory thyroid tissue, have been reported in a large number of rats following prolonged administration of thiourea. Although these "new-growths" frequently occurred coincidentally with tumors in the thyroid gland, little similarity existed between neoplasms in the two sites. They may be derived from portions of ultimobranchial tissue carried into the thymus during embryonic development, or even from hyperplastic foci of thymic epithelium. Such nodules probably develop as compensatory aberrant goiters following excessive thyrotropic hormone stimulation caused by the goitrogen. In the rat, at least, they suggest that the potentiality for ectopic thyroid tissue in the mediastinum is considerable. Similar manifestations, less marked in character but receptive to radioactive iodine, develop in thymus tissue following extirpation of the thyroid gland; and these indicate that the thymus may be a source of thyroid hormone following thyroidectomy, owing to excessive stimulation by thyrotropic hormone. There is evidence too that the ultimobranchial bodies can differentiate into thymus tissue, and possibly even parathyroid tissue, as suggested by experimentation.

In the rat, a useful animal for experimentation, ultimobranchial cysts occasionally develop spontaneously. They are most often seen in atrophic glands, in females, and in thyroids of older animals. As in sheep, the basal cell layers of these squamous cysts also produce gland-like cells. In the presence of hyperplastic parenchyma, these cells appear to transform into essentially typical thyroid tissue. In atrophic glands, these new cells sometimes form vesicles which resemble mucous acini, and staining by special method indicates a mucus reaction. Following either this cell outgrowth or separation from the basal cell component of these stratified epithelial cysts, thin-walled remnants of this epithelium persist. These represent largely only the more superficial or "suprabasal cell" layers of the original cyst. Such densely staining, thin-walled cysts characterize thyroid glands in old animals. Although these keratinizing cells are undergoing degeneration, some presumably remain temporarily viable. They are often contiguous to spontaneously developing cystadenoma; and the origin of some such neoplasms in the thyroid was assigned to these degenerating cystic masses in the past.

A number of factors (intrinsic and extrinsic) can modify ultimobranchial tissue in the rat. Ultimobranchial tissue here normally becomes indistinguishably transformed into thyroid-like tissue during embryonic development. At birth, it usually cannot be recognized as an entity separate from surrounding areas of the thyroid. According to some investigators, a similar situation holds for the development of the human thyroid gland.

In *young* rats, ultimobranchial tissue apparently functions as thyroid tissue, for storage of iodine following radioautograph techniques shows it to be little different from other thyroid areas. But this is not true when it has become cystic as in older animals. Ultimobranchial tissue is "labile," and, depending upon the age of the animal and the species, it may be represented either as a squamous cell or secretory cell type of metaplasia. In mice, for example, it is common to find mucus-secreting cysts, which may also possess cilia. Gorbman (1947) has noted that in mice these atypical follicles or cysts do not store iodine. It seems clear that ultimobranchial tissue may react differently in various species.

In view of the similarity of rat thyroid embryology to that of man, many studies have been undertaken to clarify the factors that may influence this thyroid tissue component.

Apparently, prolonged administration of a goitrogen, such as propylthiouracil, will induce ultimobranchial tissue to undergo neoplasia, but adjacent areas of thyroid tissue may also be involved, and there may even be multiple nodules. Ultimobranchial tissue is typically found at this site. Squamous cell metaplasia has also been associated with repeated injection of estrogenic hormones; but it is not certain whether these hormones induce this type of lesion in the thyroid or augment a pre-existing metaplastic site. Nevertheless, the position of these lesions is also consistent for ultimobranchial tissue. Prolonged administration of the carcinogen, methylcholanthrene also appears to be responsible for ultimobranchial tissue metaplasia, especially in mice; this may be due to the reputed estrogenic action of certain carcinogens.

Thyroids from vitamin A deficient rats *invariably* show sites of keratinizing, cystic metaplasia. Such lesions are usually in the center of both lateral lobes. Thus, they develop in the precise site occupied by the ultimobranchial body and are *predictable*. Although these changes are reversible following vitamin A therapy, certain remaining cells or proliferations from these ultimobranchial cysts may persist indefinitely after periods of repair and subsequently become tumorigenic. Chronic vitamin A deficiency will induce these lesions in animals of all ages, and this has been the basis for many experimental studies designed to determine the role of epithelial metaplasia in the evolution of certain neoplasms. It is of particular interest to note that squamous cysts arising through metaplasia can be markedly augmented by estrogenic hormones, especially during periods of reparation.

Feeding the goitrogen, allylthiourea, and the carcinogen, 2-acetyl-aminofluorene, simultaneously, during chronic vitamin A deficiency, in rats has resulted in the development of two rather common types of thyroid neoplasms. These can be attributed to the ultimobranchial body. One kind of experimental tumor (a solid, anaplastic or undifferentiated type) can be traced back to its source, the basal or germinative cell layers in stratified squamous epithelium, of these induced metaplastic cysts. It develops precociously and can become highly malignant. Another type (cystadenoma or cystadenocarcinoma) can be related to the so-called "*suprabasal cell*" component of these persisting epithelial lesions. These tumors have been interpreted as arising from certain densely staining cells which have become *re-vitalized* while undergoing degeneration in the wall of these squamous cell cysts. Since they represent cells that were also "differentiating toward keratinization," some viability is presumed to be retained. Under the conditions of this experiment, these tumors required a prolonged period of time for development. Both types of neoplasms may occur in the same thyroid gland.

Few investigators have described morphological precursors for thyroid neoplasms. Thyroid-incorporated ultimobranchial tissue, however, can be considered one source of tumorigenesis within the mammalian thyroid gland. Experimentally, this tissue may be "manipulated" in various ways during chronic vitamin A deficiency to demonstrate the origin and sequence of development of certain thyroid tumors.

Finally, in order to better illustrate foci of incipient tumor formation (especially the cystadenoma) which may be related to ultimobranchial tissue metaplasia, young rats (after an initial period of vitamin A deficiency) were alternately fed a normal (reparative) diet for seven days, followed by a vitamin A deficient diet for 21 days. This alternation was continued for six to eight months, during which time the animals received thiourea continuously.

Thyroid tumors stimulated by goitrogens *alone* were rarely associated with ultimobranchial cysts. Such tumors are usually of the cystadenoma type and, in these "morphologically activated" thyroid glands, they are often *incompatible* with "vestiges" suggestive of origin. Nevertheless, animals subjected to recurring periods of vitamin A deficiency *during* goitrogenesis frequently showed neoplasms contiguous to cystic remnants of ultimobranchial tissue. During this periodic epithelial repair, basal cells become separated away from cyst walls, and apparently contribute to the thyroid parenchyma (basal cells may require the action of a carcinogen for tumorigenesis). Consequently, in some instances, these cystic tumors could be *traced* directly in sections to "partially degenerated," yet differentiating, suprabasal cells which persisted in remnants of this stratified squamous epithelium.

The experimental evidence acquired here substantiates the concept that certain cells may become *re-vitalized* during degeneration and subsequently become stimulated to develop in an aberrant manner. Here, derived from ultimobranchial tissue within the thyroid gland, these cells formed *cystadenomata* under the influence of thyrotropin.

8

UROHYPOPHYSEAL GLAND

That peculiar morphologic elements suggestive of certain endocrine activities are present at the caudal aspect of the fish spinal cord is by no means a new finding; however, biologists have largely ignored their presence. As early as 1827 Weber described clearly the existence of a warty appendix ("Knoten") at the end of the spinal cord proper of the carp, with additional remarks that similar structures were present in other kinds of fishes such as the catfish and the burbot. According to this classic author, "Zwar ist die Endigung des Ruckenmarks in einen Knoten schon langst bekannt." Later authors have cited Arsaky (1813) and Serres (1826) as preceding Weber in the exploration of the unusual terminal morphology of the spinal cord. Early observations were confirmed by several subsequent anatomic studies, and, beginning with Rauber (1877), histologic investigation was pursued, most significantly by Verne (1914) and by Favaro (1926). Verne called the terminal swelling "renflement caudal," and Favaro designated it as "ipofisi caudale" (hypophysis caudalis). It is of particular interest that both authors, being impressed with the analogy in terms of gross histologic pattern, but without taking into account characteristic features of the respective nervous organizations, arrived at essentially the same view that the structure in question was glandular in nature and comparable to the posterior pituitary, as well as to the epiphysis, in the diencephalic region.

Another important approach was made in the United States in studies of a specialized type of cell in the posterior end of the fish spinal cord. Speidel (1919, 1922), having confirmed Dahlgren's discovery (1914) of enormous cells of peculiar appearance in the posterior part of the spinal cord of skates, worked out details of their cytology and extended research to other kinds of elasmobranchs as well as to a number of teleosts. From these studies it became evident that a morphologically discrete material was elaborated by these cells; Scharrer and Scharrer (1945) refer to Speidel's work as the earliest demonstration of a neurosecretory phenomenon. Unfortunately, Speidel's investigations were not specifically concerned with the neural relations of the secretory cells, and the fate of the secretory material was left undetermined; however, Speidel did report that, in the flounder, the spinal cord is provided with a dorsally overlapped "terminal enlargement" and that the secretory cells are most numerous and of greatest size in close proximity to this structure. This suggested at least a close spatial relationship between the Dahlgren cells and the characteristic terminal body extensively studied by European workers.

Concept of a Caudal Neurosecretory System

Recently, the older observations, integrated with newer findings, have been reinterpreted. The significant morphologic elements described above became recognized as the principal components of a definite neurosecretory system, for which the term "caudal neurosecretory system" was proposed.

This development was based on a chance observation, in the caudal region of the spinal cord of the Japanese eel (*Anguilla japonica*), of a remarkable nervous organization represented by a number of large and small secretory neurons and their secretion-bearing axons with enlarged bulbous endings. In routine histologic preparations, such as those stained with Mallory's, Mann's or Masson's technics, the neuron cell bodies show pronounced unclear polymorphism, changes in basophilia in the cytoplasmic periphery as well as around the nucleus, and granules or droplets showing affinity for acid fuchsin, eosin, iron-hematoxylin, etc. Affinity of the cytoplasmic granules for acid fuchsin is much more pronounced in histologic sections pretreated with alkaline thioglycollate. The cell bodies are located in a well-vascularized area and are often in direct contact with capillaries. Along their entire extent the axons that run posteriorly to terminate in enlarged bulbous endings show the same tinctorial response as that of the cytoplasmic secretion, indicating apparent continuity of the stainable material between the site of origin and the axon. The end-bulbs filled with colloid showing the same stainability as that of the secretory material are adjacent to capillaries running along the ventral and lateral aspects of the terminal region of the spinal cord. Here the cord tissue slightly bulges out because of

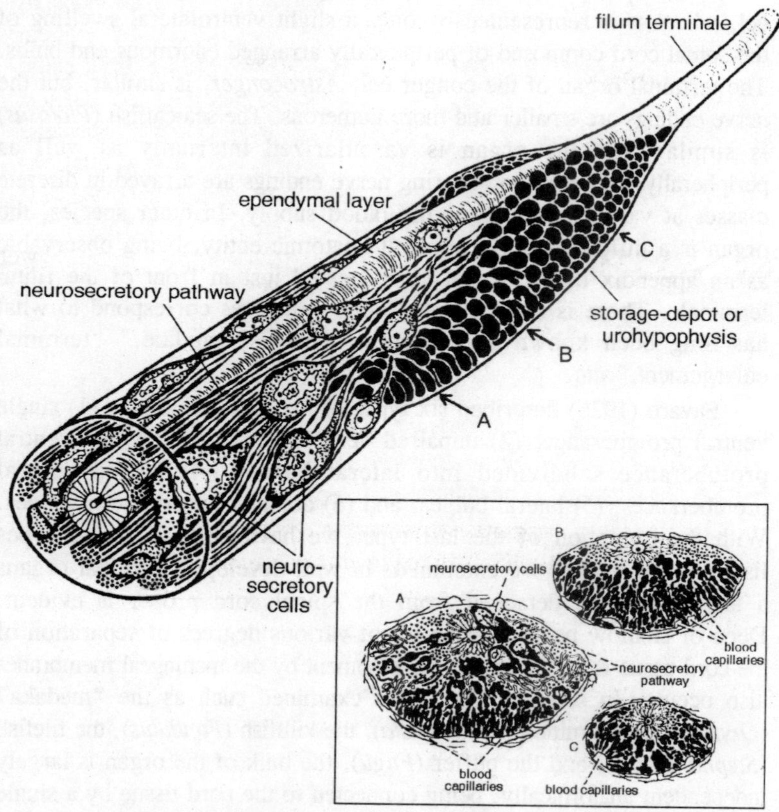

Fig. 8.1. Pattern of arrangement of the caudal neurosecretory system of the eel.

the massive aggregation of nerve endings. Occasionally large or small droplets, resembling the colloid in the end-bulbs, are found within the lumen of contiguous blood vessels. Thus, the characteristic neuron in the caudal spinal cord appears to be involved in production of a kind of secretory material, its axon serving as the neurosecretory pathway, and the axonal end-bulb responsible for storage and release of the secretory material.

With this background of information, histologic research was extended to nearly 100 species of fresh-water and marine teleosts. This survey demonstrated that in all teleost fish so far examined the characteristic terminal nervous system is developed on the same basic plan as that seen in the eel. However, there are variations in morphologic pattern, particularly in the relation of the secretion-bearing nerve endings to the circulatory system.

The simplest form of the storage-depot organization is seen in the eel, where it is represented by only a slight ventrolateral swelling of the spinal cord composed of peripherally arranged enormous end-bulbs. The terminal organ of the conger eel, *Astroconger*, is similar, but the nerve endings are smaller and more numerous. The sea-catfish (*Plotosus*) is similar, but the organ is vascularized internally as well as peripherally, and secretion-bearing nerve endings are arrayed in discrete masses at various sites along the blood supply. In other species, the organ is a differentiated prominent anatomic entity, being observable as an appendix of the spinal cord located just in front of the filum terminale. There is no doubt that such structures correspond to what has long been known as "Knoten," "ipofisi caudale," "terminal enlargement," etc.

Favaro (1926) described six types of the terminal organ: (1) single ventral protuberance, (2) unpaired ventrolateral swelling, (3) ventral protuberance subdivided into lateral halves, (4) paired ventral protuberances, (5) lateral bulges, and (6) dorsally-united lateral bulges. With the exception of the last type, we have seen all of the types listed in the species we examined. In well-developed terminal organs a tendency to be detached from the spinal cord proper is evident. Deep or shallow notches bring about various degrees of separation of the cord tissue from the organ; envelopment by the meningeal membranes also occurs. In some of the fishes examined such as the "medaka" (*Oryzias*), the topminnow (*Gambusia*), the killifish (*Fundulus*), the filefish (*Stephanolepis*), and the puffer (*Fugu*), the bulk of the organ is largely independent anatomically, being connected to the cord tissue by a single stalk or by a pair of stalks composed of all the neurosecretory axons. In his comparative study of the mammalian neurohypophysis, Hanstrom (1954) pointed out a parallelism between the evolution of the organ and the degree of facilitation of contact of the hypothalamic neurosecretory fibers with the circulatory system. This appears to be the case also with the terminal organization of the caudal neurosecretory system. The development of the blood supply and of terminal ramifications of the secretion-bearing nerve fibers parallel the tendency to be separate from the spinal cord and may be related to the increase of efficiency of the storage and release function.

Local accumulation of secretory material at various spots along the axon have been seen. This corresponds to the characteristic beaded appearance frequently noted in the neurosecretory fiber tracts in many kinds of animals. Also, a variety of peculiar structures resembling the

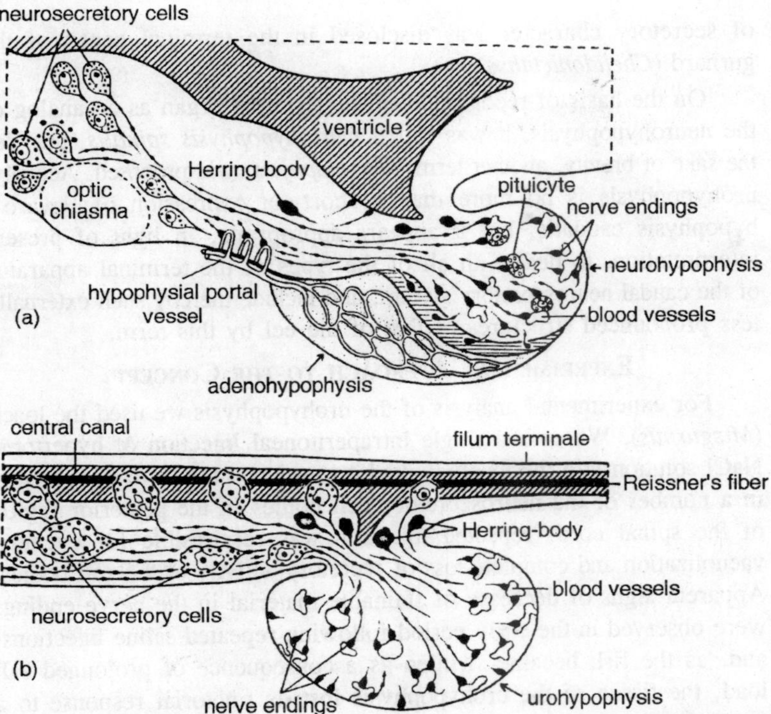

Fig. 8.2. Comparison of the pattern of organization of the caudal neurosecretory system (b) with that of the hypothalamo-hypophysial system.

so-called Herring bodies in the hypothalamo-hypophysial system is found distributed mainly at the dorsal aspect of the terminal organ. The nature of such figures is yet unknown, but some of them suggest either a degenerative change, or a kind of holocrine activity of the posteriorly distributed small neurosecretory cells.

Thus, the terminal organ of the caudal neurosecretory system is reminiscent of the neural lobe of the pituitary both in shape and basic histologic pattern. It conforms to the current concept of neurosecretion. In connection with this similarity, it is of interest to observe the presence of a short outpocketing of the central canal at the dorsal side of the terminal organ in the catfish (*Parasilurus*) and the saury (*Cololabis*), which bears some resemblance to the infundibular recess of the third ventricle near the posterior pituitary, generally, the glial elements included in the tissue of the terminal organ are not specifically differentiated into such characteristic cells as the parenchymatous pituicytes of the neurohypophysis. However, a kind of cell suggestive of secretory character was disclosed in the terminal organ of the gurnard (*Chelidonichthys*).

On the basis of recognition of the terminal organ as an analog of the neurohypophysis, it was called *neurohypophysis spinalis*, but, for the sake of brevity, another term, *urohypophysis*, was proposed. Actually, urohypophysis is no more than a short-cut expression of Favaro's hypophysis caudalis, but it appears appropriate, in light of present interpretation, to deal with all of the types of the terminal apparatus of the caudal neurosecretory system, and include thereby such externally less pronounced structures as that of the eel by this term.

Experimental Approach to the Concept

For experimental analysis of the urohypophysis we used the loach (*Misgurnus*). Whereas a single intraperitoneal injection of hypertonic NaCl solution into the loach elicited a transient state of hypersecretion in a number of the neurosecretory cell bodies in the posterior region of the spinal cord, repeated injection was responsible for eventual vacuolization and complete loss of secretory activity of the cell bodies. Apparent signs of decrease of stainable material in the nerve endings were observed in the early period following repeated saline injections, and, as the fish became swollen as a consequence of prolonged salt load, the tissue of the urohypophysis lost its tinctorial response to a great extent.

When the spinal cord was transected in front of the urohypophysis and the fish was subjected to repeated salt loading, these events occurred: At first a rapid depletion of stainable material took place in the urohypophysis, followed by a state of increased secretory activity in the small cells posterior to the site of operation. In rough parallel to the increase of activity of the cells in question, stainability of the nerve terminals was partially restored, suggesting that the accelerated release of secretory material from the nerve endings became counterbalanced by increased supply of newly formed secretion from the neighbouring small cells. Nevertheless, both the small cells and the urohypophysis eventually became depleted in the later phase of profound impairment of osmoregulation. On the other hand, the anteriorly distributed large cells, whose axonal connection with the urohypophysis was disrupted, began to show increased secretory activity after most of the posterior small cells attained their peak of activity. Occasionally, a marked picture of accumulation of secretory material was observed in the fiber tracts arising from the large cell bodies. This was particularly pronounced proximally to the level of operation,

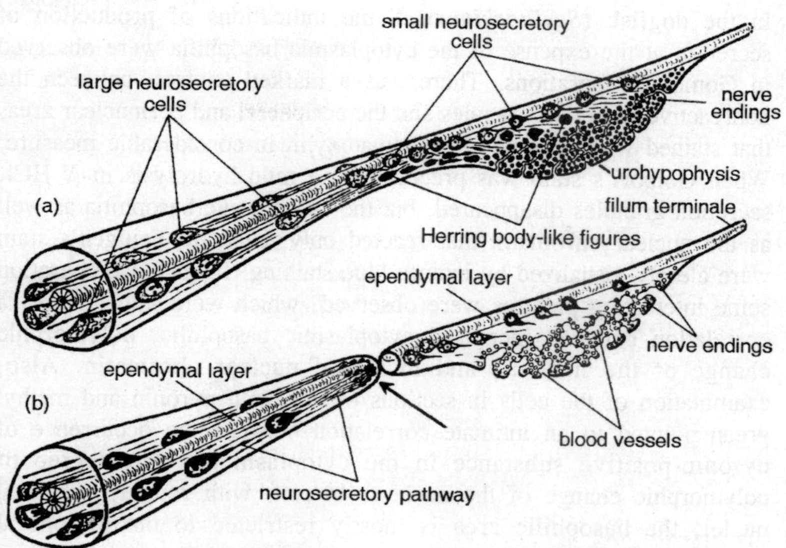

Fig. 8.3. Diagrammatic representation of the result of experiments on the loach. (a) Average appearance of the caudal neurosecretory system in fish. (b) Situation after repeated injection of hypertonic saline in fish with the spinal cord transected.

where secretion-bearing axons developed enlarged terminal swellings. This unusual accumulation of secretion of the large cells was brought to an end as signs of edema became apparent.

These observations indicate that the neurosecretory cells in the caudal spinal cord and the nerve terminals in the urohypophysis are significantly correlated so as to work in unity in response to the same physiologic demand. As in the well-explored hypothalamo-hypophysial system of vertebrates, as well as analogous organizations in crustaceans, insects, etc., peripheral movement of neurosecretory material by way of the axoplasm also occurs in the caudal neurosecretory system.

Some Cytologic Aspects of Secretion-production

The enormous caudal neurosecretory cells of the eel provide a suitable object for the cytology of secretory activity. As mentioned above, secretory material originating in the cells stains with acid fuchsin, etc., but unlike that elaborated by hypothalamic neurosecretory cells, it does not show affinity for chrome alum hematoxylin, taking phloxine pronouncedly in preparations stained after Gomori's technic. The same tinctorial response to Gomori's stain was reported by Scharrer and Scharrer (1954b) for the secretory material of the Dahlgren cell in the dogfish (*Scylliorhinus*). Some indications of production of secretion at the expense of the cytoplasmic basophilia were observed in Gomori preparations. There was a marked contrast between the nonreactive secretion granules and the peripheral and perinuclear areas that stained with chrome alum hematoxylin in considerable measure. When Gomori's stain was preceded by a mild hydrolysis in *N* HCl, secretion granules disappeared, but the cytoplasmic basophilia as well as the nuclear chromatin that reacted only feebly to Feulgen's stain were clearly visualized by intense blue staining. In such a preparation some interesting pictures were observed, which were indicative of a correlation between increased cytoplasmic basophilia, polymorphic change of the nucleus, and increased nuclear chromatin. Also, examination of the cells in sections treated with pyronin and methyl green pointed to an intimate correlation between the occurrence of pyronin-positive substance in the cytoplasm and the degree of polymorphic change of the nucleus. In cells with roughly spherical nuclei, the basophilic area is mostly restricted to the peripheral cytoplasm, but, in cells showing pronounced nuclear polymorphism, significant amounts of basophilic substance are also found adjacent to the nucleus. Generally, the perinuclear basophilia is located at large or small depressions in the nuclear membrane, being practically absent along smooth and convex portions of the nucleus. There are cells showing extremely lobate small nuclei with little or no pyronin-positive substance in either peripheral or perinuclear areas.

It appeared significant that the nucleolus, which stained intensely with pyronin, was occasionally much elongated and was in contact with the nuclear membrane. This was demonstrated also by sections impregnated moderately with protargol and counterstained with pyronin, in which the nucleoli appeared as double-structured figures consisting of an argentaffin interior and a pyronin-positive exterior. Of the elongated nucleolus, the slender apical portion that extended to the nuclear membrane was mainly represented by the pyronin-positive material. The argentaffin internal material was concentrated in the basal enlargement. By use of Hamazaki's so-called KES technic (1951), it was observed that a small aggregate of reaction-positive granules (purple-stained) was located just opposite to the attachment point of the apex of a given elongated nucleolus at the nuclear membrane. Also, seemingly spherical figures rimmed with the stainable substance were found attached to the surface of the nuclear membrane and lying in shallow depressions of the latter. Hamazaki believes that his technic stains various derivatives of the nucleic acids that cannot be observed by means of conventional histochemical methods. Whether or not such an opinion is tenable, the present finding suggests at least an intimate relationship between the nucleolar material and a certain cytoplasmic material. Although it is far from established, such a line of observation implies a possible functional link between nuclear activity and cytoplasmic basophilia in the synthesis of secretory product.

Electron Microscopy of Secretion

A recent electron-microscope study supplemented the light microscopic observation of secretory material of the caudal neurosecretory system. Within the neuron cell bodies in the eel, spherules and ellipsoids of high electron density occur, measuring approximately 1000-2500 Å in diameter. Those located at the axon hillock as well as at the proximal portion of the axon are more or less enlarged and of somewhat irregular contour, as compared with the granules at the interior of the perikaryon. As one proceeds along the axon posteriorly, large-scaled deformation of the secretion is encountered. Unlike the particles observed in the neuron cell body, those in the distal portion of the axon appear enlarged, measuring approximately 2500-4200 Å in diameter. Such enlarged particles may be of uniformly high or low electron densities; they may be of a vesicular nature, being of high density externally and relatively low density internally; some are suggestive of coalescence of different particles. Finally, within the end-bulb in the urohypophysis, secretory granules are markedly uniform, of a much diminished size of approximately 500 Å in diameter. Most of these fine granules show low electron density, but intermingled with them are granules of similar size, which are comparatively electron-dense. It is apparent that packets of such minute granules represent the stainable colloid deposit in the nerve terminals observed with light microscopy. Thus, it appears likely that the secretory material synthesized within the neuron soma becomes altered in its physicochemical properties in the course of migration by way of the axoplasm and is reduced to uniform particles by the time it reaches the nerve ending. A certain degree of "maturation" might be ascribed to the observed modification of the secretory material, although no further information of significance has been obtained in this respect.

Association of Zinc with Secretion

One of the significant properties of the secretory material of the caudal neurosecretory system in the eel is its association with zinc. Supravital staining with alkaline dithizone solution selectively demonstrates the neurosecretory system, showing specifically stained secretory material which corresponds to the picture observed in routine histologic preparations. Polarographic determination of zinc in different portions of the spinal cord showed principal concentration of the element in the caudal portion, which includes the bulk of the neurosecretory system. Measurements of fractions resulting from extraction with water, ethanol, acetone, chloroform, or petroleum ether provided data indicating that, although both soluble and insoluble forms of zinc are contained in large amounts in the caudal spinal cord, it is the former that is especially abundant, amounting to several times the quantity detected in anterior portions of the spinal cord. When the caudal spinal cord was fixed in ethanol for histochemical demonstration of zinc in sections, a loss of the tinctorial response occurred both in the axons and in the nerve endings. However, a positive response remained in sonic of the perikarya, with reactive granules or droplets simulating the secretory material to some extent. Since ordinary histologic preparations of ethanol-fixed tissues do not show any residual secretory material throughout the entire caudal neurosecretory system, the presence of zinc-reactive material unaffected by ethanol does not seem to indicate specific insolubility in ethanol of the secretory material itself at the site of origin. Rather, this observation suggests that the secretory material within the perikaryon, unlike that in the axon and in the end-bulb, exists in combination with ethanol-insoluble zinc compounds. In this respect, the electron-microscopic observation of changes during the course of peripheral migration would appear to be of particular interest.

FUNCTIONAL SIGNIFICANCE

Possibility of Participation in Sodium Exchange

The result of experimental salt loading in the loach stimulated an inquiry into the possibility of production by the caudal neurosecretory system of a hormonal principle concerned directly or indirectly with osmoregulation. The small cyprinodont "medaka" (*Oryzias*) exhibited some signs of disturbance in adaptive sodium-regulation, after removal of the part of the spinal cord lying posterior to the dorsal fin, or after transection of the cord just anterior to the base of the fin. In contrast to virtually no change in potassium-content of the body (excluding the viscera), total sodium-content was increased in comparison to that of the intact animal, after repeated transference of the operated fish from fresh water to physiologic saline and thence to half-concentrated artificial sea water, and in reversed direction. Intraperitoneal injections of homogenate or crude extract of eel caudal spinal cord were followed by a decrease of total sodium-content of the operated fish, when the recipient was preadapted to isotonic physiologic saline or hypertonic 50 per cent artificial sea water. On the contrary, an appreciable increase in sodium-content was observed as the result of such injections when the recipient had been kept in fresh water.

These observations suggest the possibility of a relation between the caudal neurosecretory system and certain target organ(s) responsible for excretion and uptake of sodium. Concerning the endocrine control of adaptive osmoregulation in fishes, it is not possible to synthesize the various views and provisional conclusions presented to date and thus to delineate at least the outline of osmoregulatory coordination. Much more research is needed to obtain any clear-cut picture of the problem. Future work might consider the possibilities raised by the experiments reported above.

Relation to Gas Metabolism

The chance observation that the goldfish (*Carassius*) loses buoyaney on removal of the tail posterior to the base of the anal fin made possible a fruitful line of research. Intraperitoneal injection of a crude extract of eel caudal spinal cord results in a marked increase of buoyancy, which appears as a kind of forced floating of the tailless fish. By measuring the change of buoyancy of the fish body as a whole according to the principle of cartesian diver manometry, events resulting from injection of effective extracts can be easily followed. In practice, the test goldfish is deprived not only of its tail but also of all fins with the exception of the dorsal one, in order to eliminate as far as possible spontaneous movements of the body. Postoperative mortality is less than 5 per cent for at least 7 days. Usually, experiments are carried out on the eighth day, when the fish shows the characteristic state of continuous submergence.

By injection of a crude eel extract, two successive phases of increase of buoyancy are evoked. That the two responses are dependent on different substances was learned from experiments involving injection of various eluates of a paper-chromatogram of a crude extract developed with *n*-butanol-acetic acid-water (40 : 10 : 50). Of the eluates obtained from four conveniently separated regions of the chromatogram as visualized by staining with bromphenol blue, ninhydrin, and dithizone, the eluate prepared from the topmost section elicited the first response, which attained its maximum in approximately 120 minutes at 20°C, whereas the eluate of a lower section was responsible for the second

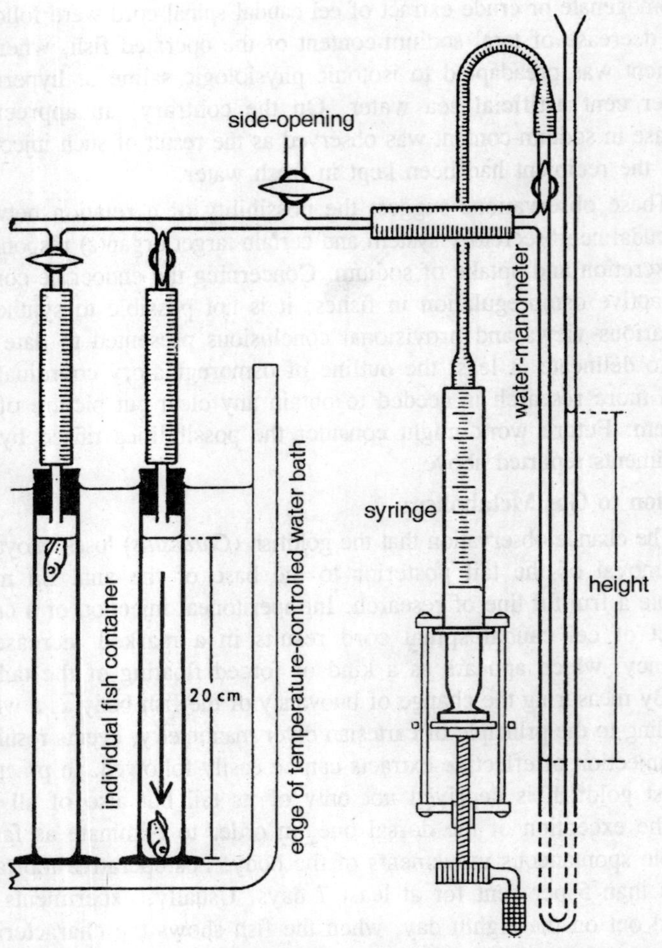

Fig. 8.4. Apparatus for measuring change in buoyancy of tailless and finless goldfish according to the principle of certesian diver manometry.

response initiated about 150 minutes after injection. It is of interest that the test goldfish exhibited some signs of respiratory disturbance (diminution of amplitude and increase in frequency of the opercular movement) following injection of either the original crude extract, or the eluate of the topmost section of the paper chromatogram; no external sign of respiratory impairment was observed for at least 4 or 5 hours following injection of the eluate of the lower effective section. The phenomenon seems to be related to the occurrence of dithizone-positive bands in the topmost section, which appear to be identified with the presence of zinc, in view of the observed abundance of the soluble

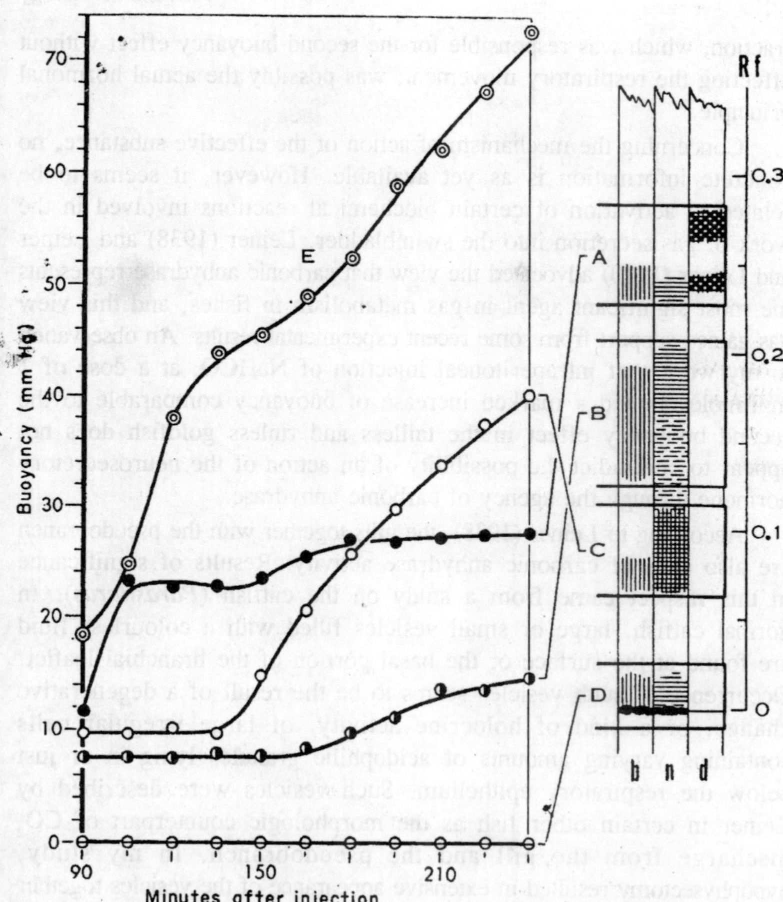

Fig. 8.5. Effects of injection of fractions obtained by paper-chromatographic development of a crude extract of eel caudal spinal cord, in comparison with that of the original extract (curve E). b, bromphenol blue-treatment; n, ninhydrin-treatment; d, dithizone-treatment.

form of this element in the caudal spinal cord of the eel. Paper chromatography of $ZnSO_4$ has shown a dithizone-positive band at Rf 0.25 under the same experimental conditions. Intraperitoneal injection of $ZnSO_4$ at doses in the range of 0.1-0.01 millimoles simulates the result of injection of the eluate of the topmost section both in regard to buoyancy and respiratory effects. Because of this similarity, the observed activity of the topmost section of the paper chromatographic development of the crude extract was considered to be due to its zinc ingredient; it was conjectured that a substance included in the lower fraction, which was responsible for the second buoyancy effect without affecting the respiratory movement, was possibly the actual hormonal principle.

Concerning the mechanism of action of the effective substance, no concrete information is as yet available. However, it seems to be related to activation of certain biochemical reactions involved in the work of gas secretion into the swimbladder. Leiner (1938) and Leiner and Leiner (1940) advocated the view that carbonic anhydrase represents the most significant agent in gas metabolism in fishes, and this view has gained support from some recent experimental results. An observation in my work that intraperitoneal injection of $NaHCO_3$ at a dose of 1 millimole elicited a marked increase of buoyancy comparable to the second buoyancy effect in the tailless and finless goldfish does not appear to contradict the possibility of an action of the neurosecretory hormone through the agency of carbonic anhydrase.

According to Leiner (1938), the gills together with the pseudobranch are also sites of carbonic anhydrase activity. Results of significance in this respect came from a study on the catfish (*Parasilurus*). In normal catfish, large or small vesicles filled with a colourless fluid are found at the surface of the basal portion of the branchial leaflet. Occurrence of such vesicles seems to be the result of a degenerative change, or a kind of holocrine activity, of large irregular cells containing varying amounts of acidophilic granules lying in or just below the respiratory epithelium. Such vesicles were described by Leiner in certain other fish as the morphologic counterpart of CO_2 discharge from the gill and the pseudobranch. In my study, hypophysectomy resulted in extensive appearance of the vesicles together with the characteristic degenerating cells, whereas these vesicles and relatively intact large cells wholly disappeared after removal of the caudal spinal cord or after transection of the cord at levels anterior to the site of the caudal neurosecretory system. The effect of surgical operations on the caudal spinal cord was reproduced to some extent by intramuscular implantation of excess pituitaries: on the other hand, subcutaneous injection of the eluate of the significant lower section of the paper chromatogram of the extract of eel caudal spinal cord imitated the effect of hypophysectomy. Application of the active fraction of the eel extract to the branchial leaves maintained *in vitro* also resulted in pronounced appearance of the characteristic histologic picture. Furthermore, faradic stimulation of the spinal cord, injection of acetylcholine, adrenalin, noradrenalin, pilocarpine or physostigmine, and injection of concentrated NaCl solution were equally responsible for an increased appearance of the special cells and vesicles.

Such observations may be interpreted in accordance with Leiner's contention, or may be used in support of another kind of explanation. The alternative explanation is based on interpretation of the characteristic changeable cell as the so-called "chloride cell," in line with the opinion of Keys (1931) and Keys and Willmer (1932) and substantiated by Copeland (1948, 1950). These authors insisted that the large acidophilic cell in the gill is charged with dual activities, being responsible for excretion of chloride in a hypertonic medium and for absorption in a hypotonic medium. Such a view would provide a pertinent rationale for the changes of total Na-content observed under the influence of the crude extract of the caudal spinal cord. Nevertheless, it must be stated that the functional significance of the "chloride cell" is still being disputed. In fact, the recent work of Burden (1956) has shown no obvious change of the cell in the killifish (*Fundulus*), despite extensive loss of serum chloride in fresh water after hypophysectomy. Burden is inclined to ascribe the observed profound impairment of chloride-exchange to a loss of the protective role of the mucus, stating that the mucous cells in the gills underwent extensive atrophy in parallel with disturbances in osmoregulation. In my study, the mucous cells in the gills and in the skin of the catfish were brought to hyperactivity (accelerated holocrine activity) by various experimental treatments, but none of the latter seems to be correlated specifically with the osmoregulatory phenomenon.

In general, many problems remain unsettled in the analysis of the coordinating mechanism(s) concerned with the phenomena of Na-exchange, gas-metabolism, and gill morphology. However, if one emphasizes the significance of carbonic anhydrase, the activity of which is known to be related also to the so-called chloride shift, a common basic mechanism underlying the three kinds of phenomena might be imagined.

ISOLATION OF THE EFFECTIVE FRACTION

Very recently an attempt at isolation of the caudal neurosecretory hormone was made, employing the buoyancy effect as a valid reference for assay. Preliminary information derived from extractions of nearly 6000 isolated spinal cords of the eel, has shown that the expected hormonic substance can be separated from the fraction responsible for respiratory disturbance by precipitation with 85 per cent acetone. It is soluble in water as well as in various concentrations of ethanol, is stable in acid media but liable to lose activity in alkaline media, is dialyzable through collodion and cellophane membranes. In one experiment 2458 isolated caudal spinal cords of the eel were subjected to the following procedure.

Essentially the procedure consisted of (1) removal of lipids, (2) precipitation of the active principle with 85 per cent acetone, (3) fractionation with 45 per cent ethanol, and (4) further fractionation by means of paper-electrophoresis. On assay, the ethanol-soluble fraction designated as Fraction A was found to be exclusively active, others designated as Fractions B, C, D, and E being without observable effect. Paper-electrophoresis of Fraction A in phosphate buffer of pH 7.8 and $\mu = 0.6$ showed development of roughly 5 ninhydrin-positive bands, of which the one moving to the anode was found to include the activity. The eluate of this band (A1) was run again on paper in the same buffer but of lower ionic strength ($\mu = 0.2$), resulting in its separation into two ninhydrin-positive bands, of which the one proceeding to the anode (A2) contained most of the activity. On further paper-electrophoresis of the active eluate in acetate buffer of pH 3.0 and $\mu = 0.01$, two ninhydrin-positive bands developed showing similar mobilities in opposite directions. The one that moved to the anode (A3) was found to be effective, in contrast to the practically ineffective one moving to the cathode (A4). Paper chromatography of A 3 with a mixture of *n*-butanol-acetic acid-water (40:10:50) showed a significant

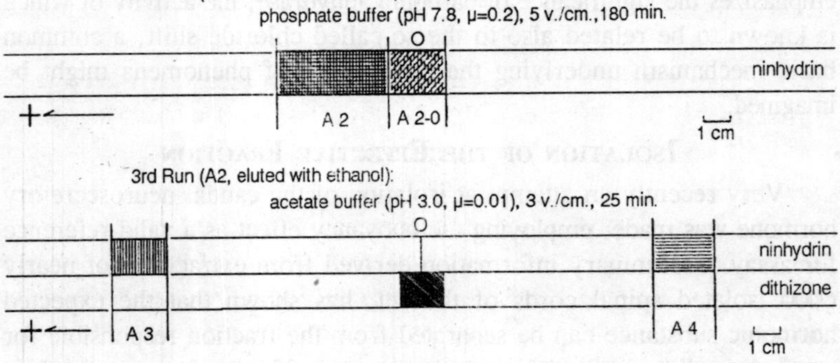

1st Run (Fraction A): phosphate buffer (pH 7.8, μ=0.5), 7 v./cm., 120 min.

O

ninhydrin

+←

A 1 A 1-0

1 cm

2nd Run (A1, eluted with water of pH 5.0): phosphate buffer (pH 7.8, μ=0.2), 5 v./cm.,180 min.

O

ninhydrin

+←

A 2 A 2-0

1 cm

3rd Run (A2, eluted with ethanol): acetate buffer (pH 3.0, μ=0.01), 3 v./cm., 25 min.

O

ninhydrin

dithizone

+← A 3 A 4

1 cm

Fig. 8.6. Paper-electrophoretic fractionation.

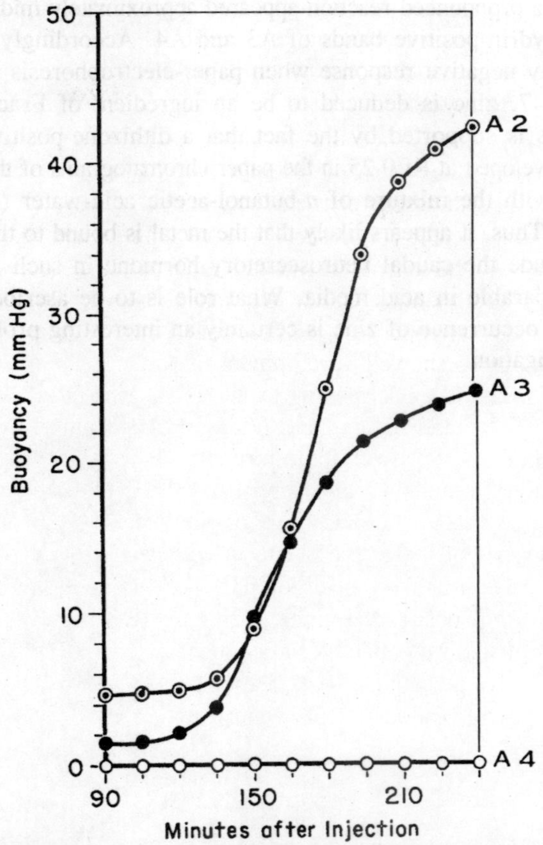

Fig. 8.7. Comparison of results of assay of fractions A2, A3, and A4.

ninhydrin-positive spot around Rf 0.13, being associated with less marked ones detected at Rf 0.07 and Rf 0.18, respectively. Further efforts to obtain an ultimately purified substance in sufficiently large amounts should be made before any attempt is made to ascertain the chemical properties of the hormone. For the time being, it can be stated that the practically isolated effective fraction appears to include polypeptide-like substances.

As regards the significant occurrence of zinc in association with the secretion of the caudal neurosecretory system of the eel, fractions obtained in the process of extraction described above provided some interesting information. In the paper-electrophoresis of Fractions A and A 1 at pH 7.8, no characteristic dithizone-positive reaction was observable. However, in the paper-electrophoresis of Fraction A2 at pH 3.0, a pronounced reaction appeared approximately midway between the ninhydrin-positive bands of A3 and A4. Accordingly, despite the seemingly negative response when paper-electrophoresis is conducted near pH 7, zinc is deduced to be an ingredient of Fractions A and A1. This is supported by the fact that a dithizone-positive band was found developed at Rf 0.25 in the paper chromatograms of these fractions treated with the mixture of *n*-butanol-acetic acid-water (40:10:50) at pH 2.5. Thus, it appears likely that the metal is bound to the complexes that include the caudal neurosecretory hormone in such a manner as to be separable in acid media. What role is to be ascribed to such a mode of occurrence of zinc is certainly an interesting problem worthy of investigation.

9

CORPUSCLES OF STANNIUS

The corpuscles of Stannius have been, from time to time, described as the adrenocortical homologue of the Teleost fishes. Despite the considerable volume of work on these enigmatic bodies, mostly in the latter part of the nineteenth century, their real function is still in doubt. In addition to the adreno-cortical hypothesis they have also been held homologous with the Mullerian duct and latterly, in a persuasive paper, as osmoregulatory organs.

This paper, whilst recalling previous work, adds some comments on the nature of these bodies, their development, histological picture, and mode of function in some of the salmonid fishes of the West coast of North America.

EMBRYOLOGY

The earliest work on the development of these bodies deserving of attention is that of Giacomini (1912, 1920, 1922), in conjunction with information on the origin of the adrenocortical tissue in salmonids and lophobranchs. In a later paper Garret (1942) describes their development in *Amia calva* and in *Salvelinus fontinalis* as well as several other Teleosts and he confirms the work of Giacomini and others.

In the Pink salmon of the Pacific coast (*Oncorhynchus gorbuscha*) the first appearance of the corpuscles takes the form of groups (two pairs) of neutrophilic cells in the dorsal aspect of the pronephric duct, readily distinguishable from the basophilic cells of the duct itself. The duct stage is followed by a period of rapid cell division producing a swelling in the wall of the duct. It is attached to the duct by a stalk for a short period and then migrates into the kidney substance. Following the stalk period, and as the body moves off into the kidney, it becomes invested by venous sinusoids but it is itself still avascular. It is situated, however, close to the dorsal aorta which lies immediately dorsal to the developing mesonephros.

As the body increases in size the venous sinusoids penetrate it, coursing between undifferentiated and rapidly dividing cell groups. As it reaches the dorsal surface of the kidney it receives an arterial supply either from the dorsal aorta direct or via a renal artery. In the Steel-head trout the bodies lie lateral to the kidney, and the ultimate site varies from species to species.

The body of the corpuscle now assumes its definitive form but no cell types are distinguishable although the cells contain many basophilic granules. In the Pink salmon two pairs are the normal complement but occasionally one pair or three corpuscles may be found. The attempt to homologize these bodies with the Mullerian duct is not a valid thesis since, as de Beer says, the proof of homology lies in the adult structure not in ontogenetic stages.

Histology

In the Pink salmon the corpuscles of Stannius are large bodies 5 × 8 × 3 mm, yellow pink in colour, deeply, although loosely, buried in the dorsal surface of the kidney. They have a dorsal hilus in which arteries and veins of vascular supply and drainage can be clearly seen. The arteries are of medium size. The gland is invested in a thin collagenous capsule which penetrates and divides it into lobes and lobules. The vascular supply follows the collagenous septae.

In the Pink salmon there appear to be two types of morphological arrangement for the cells. Peripherally the cells are arranged in cords into which the sinusoids pass, often with a marked central vessel. The cells apposed to the sinusoids are regularly arranged with their proximal borders filled with secretory granules. The nuclei, large, and ovoid to round, basophilic, and somewhat vesicular, lie below the cytoplasm distal to the sinusoid. In some of the cords the centre is filled with irregular masses of cells whose nuclei are often intensely basophilic and may be shrunken, indicative of degeneration.

Toward the centre of the gland the cells are arranged in anastomosing plates, commonly two cells thick, with the sinusoids flowing between the plates. The cells have the same disposition as those in the peripheral cords and do not exhibit any difference which may be related to function. The plates do not show the pycnotic cells of the cords.

Extensive attempts, using a variety of polychrome and other stains failed to reveal any cell types in the gland and it is suggested that only the one type of cell is present. In addition to the secretory granules the cells contain many mitochondria.

The sinusoids are lined with reticuloendothelial cells, some of which are modified into large phagocytes of the same type as the cells of von Kupfer in the liver.

Experimental Results

Samples of blood were taken by ventricular puncture from a mature population of Sockeye salmon (*Oncorhynchus nerka*). The samples were collected directly into sterile dextrose-citrate. From the same fish from which the blood was taken, corpuscles were collected into suitably diluted methanol.

The citrated plasma and the methanol preserved Stannius corpuscles were extracted by the methods of Axelrod and Zaffaroni, and Zaffaroni. Chromatographic separation was carried out on paper beside a 10-gamma standard of cortisone and hydrocortisone. An appreciable concentration of hydrocortisone can be seen in the plasma but is entirely lacking in the corpuscles. The results of the chromatograms would seem to indicate conclusively that the corpuscles are not responsible for the production of corticosteroids of the C^{21} type.

10

Pancreatic Islets

In the history of islet research the fishes play an important role. In 1846, Stannius and Brockmann described the Brockmann bodies in teleosts. An internal secretion of the pancreas, however, was not the issue of these days. About 20 years later, when Langerhans (1869) discovered the islets in the rabbit pancreas, he felt that he even had "to refrain from a hypothesis on the character and value of our cells." When again Langerhans (1873) found the islet tissue in ammocoetes, he considered it as "the (exocrine) pancreas" of this animal. After Diamare (1895) and Laguesse (1895) had shown that the Brockmann bodies contain the equivalent of the endocrine pancreas of other vertebrates, Massari (1898) demonstrated the existence of two different types of islet cells in teleosts. It was again in teleosts, where Bowie (1925) first described a third granular type of islet cell.

Using extracts from Brockmann bodies, in 1904 Diamare and Kuliabko tried to study the role of the endocrine pancreas in carbohydrate metabolism. At the same time, Rennie even treated human patients with preparations from fish islets.

After the discovery of insulin, Macleod (1922) demonstrated its islet origin by a comparison of the glycemic effects of extracts from Brockmann bodies and zymogen tissue. This was seemingly confirmed by the observation that removal of the Brockmann bodies is followed by a long-lasting hyperglycemia.

McCormick (1924) and McCormick and Noble (1925) also considered the commercial production of insulin from Brockmann bodies. In 1929, Jensen and co-workers achieved a preparation of crystalline fish insulin. But at this time, advanced extraction techniques for mammalian pancreas made it unnecessary to prepare fish insulin for therapeutic purposes. After the second world war, again this possibility was considered in Germany; and in Japan insulin from teleosts was used clinically.

In recent years, fish insulins became important in the research on biological and immunological hormone specificities, and Yalow and Berson (1964) discussed the use of fish insulin in diabetics with immunological resistance to mammalian preparations.

Various investigators also took advantage of the Brockmann bodies in studies on the biosynthesis of insulin. However, the use of mammalian islets may become more common, owing to improved microdissection techniques. Nevertheless, the Brockmann bodies might well become important in future studies on still unknown islet hormones.

There is an increasing evidence that the functions of islet hormones in fishes differ to some extent from those in mammals. This opens the possibility that we might learn by studying fishes more about the functions of the islet hormones which are less obvious in higher vertebrates. Furthermore, the apparently high species specificity of glucagon, and its common origin with secretin certainly will prompt studies on the evolution and functions of this hormone in fishes.

Phylogeny of the Endocrine Pancreas of Fishes

Occurrence and Action of Insulin in Invertebrates

Until a few years ago, little was known about the occurrence of the pancreas hormones in phylogeny. However, recent studies on insulin activities and B-cell-like elements in lower metazoons suggest that insulin (or an insulin-containing larger molecule?) already was present in the common ancestors of both protostomians and deuterostomians. It appears highly desirable to learn more about its molecular structure, especially with respect to the site of its biological activity (the spatial arrangement of the three disulfide bonds?) and its possible relation to proinsulin.

Glucagon, on the other hand, may be a late acquisition of the higher deuterostomians. Its apparent lack in the cyclostomes is in relatively good agreement with Weinstein's conclusion (1968) that glucagon was established in the mesozoic or late paleozoic era. According to his molecular-genetic calculations, glucagon and secretin, which were originally produced by one gene, became independent of each other at this time.

General Trends in Pancreas Evolution

All fishes have endocrine and exocrine pancreas tissue, yet great morphological variations and a large number of endocrine cell types are reflections of phylogenetic distances and of differing evolutionary trends. In the light of recent findings, it is possible to differentiate four types of pancreas:

1. A cyclostome type with total separation of specialized exocrine and endocrine tissue
2. A primitive gnathostome type, where, within a compact pancreas, the ductlike arrangement of the endocrine tissue resembles the early stages of human islet formation (many selachians and *Latimeria*)
3. A tetrapodlike type of more or less compact pancreas with typical islets (holocephalians, *Protopterus*, and some teleosts)
4. An actinopterygian type, with a tendency to develop a disseminated pancreas which is widely scattered throughout the body cavity and some of its organs (liver, spleen, and ovary), and with a partial separation of endocrine and exocrine tissue

However, the islets of the holocephalians differ strongly in their cytology from those of the tetrapods and show a close topographic relation to the ducts. The pancreas of *Protopterus* is (secondarily?) embedded in the intestinal wall.

Among the gnathostomes, it is possible to trace two phylogenetic trends:

1. To concentrate the endocrine cells as "islets" within a compact pancreas. This tendency is evident in some elasmobranchs and in *Latimeria*, and it is highly developed in the holocephalians, *Protopterus*, actinopterygians, and tetrapodes.
2. To split the originally compact pancreas of the actinopterygians, which finally results in a strong accumulation of islet tissue within Brockmann bodies.

Cytological Characteristics and Nomenclature of the Islet Cells

With exception of the cyclostomes, all vertebrates appear to have three functionally independent forms of islet cells: A cells, which produce glucagon; B cells, which produce insulin; and D cells, which very likely are the source of a third pancreatic hormone. Concerning the nature of the secretion of the D cells, there are presently at least two hypotheses: (1) This hormone is identical with gastrin, and (2) it is specifically concerned with the fat mobilization in the liver. The islets of many vertebrates also contain small, agranular cells which were first described by Bensley (1911) as C cells in the guinea pig. At least some fishes and amphibia have "amphiphil" cells which very likely represent intergrade stages between two types of granular islet cells. Two groups of fishes possess peculiar cell types of unknown function which have not yet been identified in other animals with certainty: The cyclostomes have gomori negative granular cells which electron microscopically contain empty vesicles. The holocephalians have many argyrophil X cells.

Unfortunately, the nomenclature of the islet cells is even more confusing than their number. In fishes, the only nomenclatural difficulty so far is offered by the use of the terms "A and D cell." Hellman, Hellerstrom and their co-workers differentiate between A_1 and A_2 cells. The A_1 cells of Hellman and Hellerstrom (1960) are identical with the D cells in the nomenclature of this article, while their A_2 cells arc here termed "A cells."

Comparative Islet Histophysiology of the Fishes

Cyclostomata

The islet organ of the cyclostomes is represented by an aggregation of small lobules which are surrounded by connective tissue. Intralobular cavities are common in ammocoetes and less abundant in adult lampreys. In *Myxine*, they are often absent in smaller specimens.

In *Petromyzon*, the endocrine tissue is located dorsally and ventrally to the intestinal tract at the transition from foregut to midgut; in the adult animal, the ventral portion also extends into the liver. In *Myxine* and *Bdellostoma*, the islet tissue surrounds the bile duct.

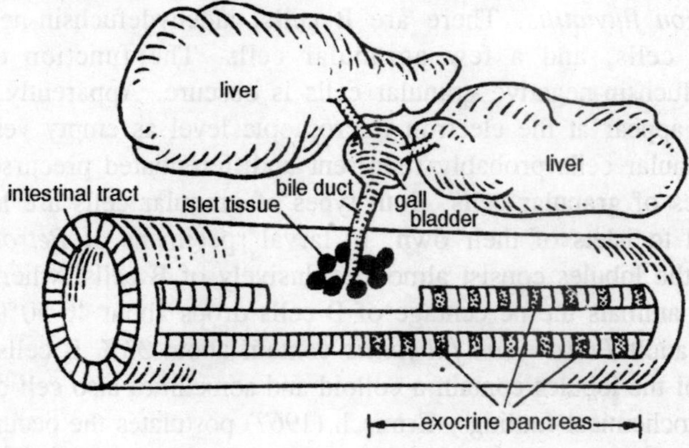

Fig. 10.1. Cyclostome type, Myxine.

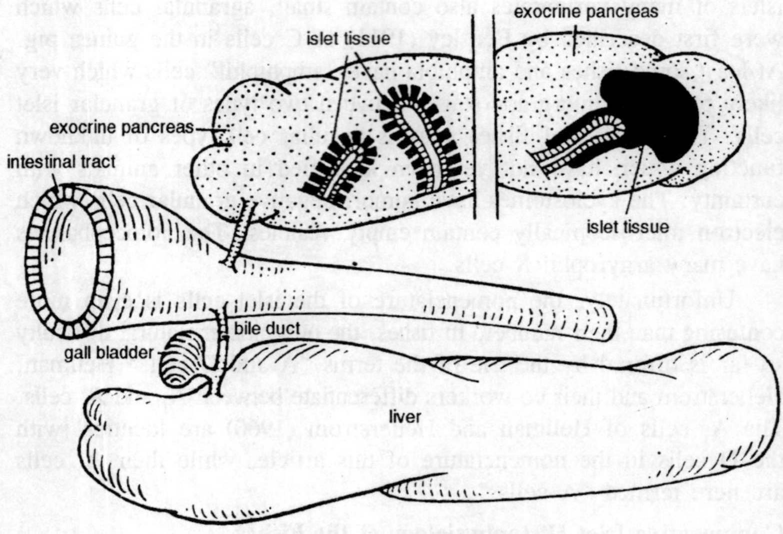

Fig. 10.2. Primitive gnathostome type.

The exocrine pancreas of the cyclostomes is separated entirely from the islet tissue. According to Schirner (1963a,b), it consists in *Myxine* and *Bdellostoma* (1) of zymogen cells in the intestinal epithelium, (2) of an intrahepatic portion, and (3) of scattered elements in the ligamentum hepatogastricum. According to Luppa and Ermisch (1967), the exocrine pancreas of *Petromyzon* and *Myxine* is represented only by specialized cells of the intestinal epithelium.

Using modern methods, Ermisch (1966) and Winbladh (1966) recently studied the endocrine pancreas of *Petromyzon planeri* and *Petromyzon fluviatilis*. There are B cells, aldehydefuchsin-negative granular cells, and a few agranular cells. The function of the aldehydefuchsin-negative granular cells is obscure. Apparently, their granules appear at the electron microscopic level as empty vesicles. The agranular cells probably represent undifferentiated precursors of both types of granular cells. Both types of granular cells are largely restricted to islets of their own. In larval specimens of *Petromyzon planeri*, the lobules consist almost exclusively of B cells, whereas in the adult animals the percentage of B cells drops about 40-50%. The islets of adult *Petromyzon fluviatilis* contain about 33% B cells. The cavities of the lobules contain a colloid and sometimes also cell debris; from histochemical findings, Ermisch (1967) postulates the occurrence of insulin; B cells may also contain colloid droplets.

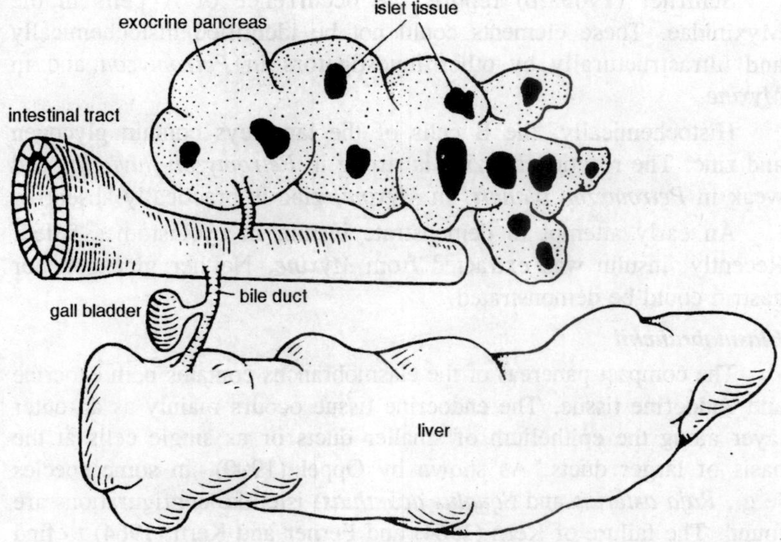

Fig. 10.3. Tetrapodlike type.

The cytology in *Myxine* is similar to *Petromyzon*. B cells make up two-thirds of the islet cells, whereas the rest consists mainly of "agranular" cells. However, ultrastructurally they contain empty vesicles and thus appear to correspond to the second type of granular cells of *Petromyzon*. The content of the lobule cavities in *Myxine* apparently is the same as in *Petromyzon*.

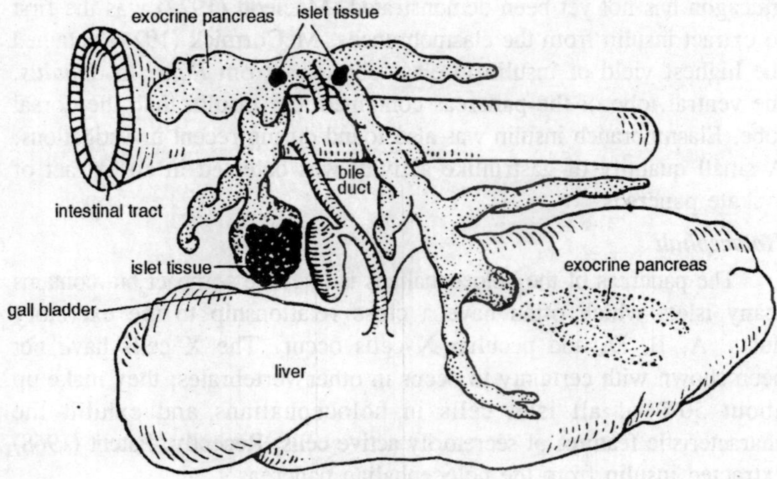

Fig. 10.4. Actinopterygian type (teleost).

Schirner (1963a,b) reports the occurrence of A cells in the Myxinidae. These elements could not be identified histochemically and ultrastructurally by other investigators in *Petromyzon* and in *Myxine*.

Histochemically, the B cells of the lampreys contain glycogen and zinc. The reaction for zinc is strong in *Petromyzon fluviatilis* and weak in *Petromyzon planeri*. In *Myxine*, zinc is practically absent.

An early attempt to demonstrate insulin in cyclostomes failed. Recently, insulin was extracted from *Myxine*. Neither glucagon nor gastrin could be demonstrated.

Elasmobranchii

The compact pancreas of the elasmobranchs contains both exocrine and endocrine tissue. The endocrine tissue occurs mainly as an outer layer along the epithelium of smaller ducts or as single cells at the basis of larger ducts. As shown by Oppel (1900), in some species (e.g., *Raja asterias* and *Squalus acanthias*) isletlike configurations are found. The failure of Kern (1964) and Ferner and Kern (1964) to find isletlike arrangements in their material possibly results from seasonal factors or the age of their animals. Howbeit, the islet tissue of the elasmobranchs shows striking similarity to the early ontogenetic stages of the islet formation in the human.

There are A, B, D, and agranular cells as well as amphiphils. This suggests the existence of insulin, glucagon, and at least one more islet hormone. Tryptophan was found histochemically in the A cells; glucagon has not yet been demonstrated. Macleod (1922) was the first to extract insulin from the elasmobranchs. McCormick (1924) obtained the highest yield of insulin in elasmobranchs from *Raja*. In *Squalus*, the ventral lobe of the pancreas contained less insulin than the dorsal lobe. Elasmobranch insulin was also found during recent investigations. A small quantity of gastrinlike activity was detected in an extract of a skate pancreas.

Holocephali

The pancreas of the holocephalians is likewise compact but contains many islets which often have a close relationship to the excretory ducts. A, B, D, and peculiar X cells occur. The X cells have not been shown with certainty to occur in other vertebrates; they make up about 50% of all islet cells in holocephalians and exhibit the characteristic features of secretorily active cells. Recently, Patent (1968) extracted insulin from the holocephalian pancreas.

Choanichthyes

Dipnoi

The pancreas of *Protopterus* lies within the dorsal wall of the gut just before its spiral portion. According to Leiner and Schmidt (1957), it contains islets with A, B, D, and C cells.

Crossopterygii

The pancreas of *Latimeria* was recently discovered by Grossner (1968); it is a compact organ at the dorsal surface of the spiral gut. The structure described as "pancreas" by Millot and Anthony (1958) consists of adipose and hemopoietic tissue. The topography of the islet tissue resembles closely that of the selachians: it occurs as a single epithelian layer along the outside of ducts, or as small buds consisting of B cells. Also, A and D cells can be identified, and two granular types of intertubular cells deserve further studies.

Actinopterygii

Based on macroscopic observations, and in accordance with Stannius, Legouis (1873) differentiates three types of pancreas among the teleosts. A compact pancreas, similar to that in most other gnathostomes; a diffuse pancreas, split into lobes that might extend to various sites of the body cavity; and a disseminated pancreas, which is scattered in small portions over the entire body cavity. As can be seen from a compilation by Oppel (1900), there are various transitions between these types: Both a compact and diffuse pancreas occur, for example, in *Pleuronectes*, and a diffuse and disseminated pancreas are both mentioned in *Gadus* and the *Cyprinidae*. Intrahepatic and also intrasplenic pancreas can be found in various species. In some cases intrasplenic and intrapancreatic or even intraovarial islets were observed. Apparently, the splitting of the pancreas can be influenced by the degree of the development of the adipose tissue. Pathological enlargement of the adipose tissue caused pancreas degeneration in hyperphagic carp.

Chondrostei

The pancreas of *Acipenser* corresponds either to the "diffuse" or to the "disseminated" type.

Holostei

The pancreas of the bowfin, *Amia*, is "disseminated" or "diffuse," and *pancreas intrahepaticum* is present. No Brockmann bodies occur, and the islets resemble those of the higher vertebrates. The islet cytology appears to be unknown. Recently, an unusual type of insulin was extracted. The pancreas of *Lepisosteus* is less disseminated than that of *Amia* and *Acipenser*.

Teleostei

Contrary to popular belief, the Brockmann bodies usually do not contain only pure islet tissue. As stressed by Baron (1934) and Bargmann (1939), in many cases they consist of a giant islet which often is separated from a more or less complete rim of exocrine tissue by a connective tissue sheet. The connective tissue capsule is not always complete, and exocrine tissue may occur even within the capsule. Thus the name "principal islet" is misleading.

In accordance with Siwe (1926), Baron (1934) considers the Brockmann bodies as direct derivatives of the dorsal epithelium of the embryonic gut. He furthermore assumes that in postembryonic life the islets in the derivatives of the ventral pancreas "anlagen" develop from ducts. This latter conclusion is supported by the observations of Dancasiu (1960). To some extent, the Brockmann bodies may correspond to the splenic portion of the avian and reptilian pancreas, which are also extremely rich in islet tissue.

It is possible to free the endocrine components of the Brockmann bodies from the other tissues. However, histological controls are prerequisite when islet material from a new species is to be collected for biochemical studies, especially since cysts of nematodes or lymphatic tissue have on appearance similar to pancreas nodules. Favourable species for the collection of islet material are *Cottus scorpius*, *Cottus quadricornis*, *Opsanus tau*, and especially *Lophius piscatorius*. The tuna, *Thunnus thynnus*, has two especially large Brockmann bodies with a different B cell content and correspondingly different amounts of insulin.

It is clear from the morphology of the teleost pancreas that investigations on the effects of "isletectomy" by removal of the Brockmann bodies must take into consideration the existence of islets at other sites. Maybe a complete removal of the islets and of all potentially insulogenic tissue is only possible in species which have a "compact" pancreas (e.g., *Anguilla vulgaris*), but then this would be pancreatectomy.

The histology of the teleost islets appears similarly complex to that of the selachians. As already shown by Bowie (1925), there are at least three forms of granular cells. Recent studies revealed A, B, and ls in all species investigated. The electron microscopic studies showed the existence of A and B cells in all teleosts investigated so far. Falkmer and Olsson (1962) found five ultrastructurally different cell types in *Cottus scorpius*. Titlbach (1966) observed A, B, and D cells; Bencosme et al. (1965) found a few granules in the D cells, while Watanabe (1960) was able to identify D granules only with the light microscopy. Like et al. (1964) observed A, B, and agranular cells and considered the latter elements as progenitors of the B cells. The observation that *Tilapia mossambica* only has B cells could not be confirmed. In some species, e.g., *Cottus scorpius* and *Scorpaena scrofa*, large numbers of agranular (C) cells are found. They are rare in *Salmo trutta*, and apparently absent in *Salvelinus leucomaenis pluvius*. Amphiphils have been described so far in *Tilapia mossambica*, *Salmo trutta*, and *Salvelinus leucomaenis pluvius*. Histochemically, the A cells show a positive reaction to tryptophan; in *Ictalurus*, they also contain glycogen. The B cells of *Opsanus tau*, *Cyprinus carpio*, and *Ictalurus nebulosus* contain glycogen. Further data on the histochemistry of the teleost islets are given by Pallet et al. (1957) and Schatzle (1958). The islets of the teleosts are rich in zinc. In *Cottus scorpius*, this material is restricted to a central islet region which contains B and D cells. In *Cottus quadricornis*, Pihl (1967) was able to localize heavy metal electron microscopically within the B granules. Various authors observed intracellular colloid droplets in A and/or B cells; although these structures were studied very thoroughly by histochemical techniques, their function is still obscure.

Since the pioneer study of Macleod (1922), more than sufficient evidence for insulin within the teleost islets has been accumulated. The occurrence of glucagon is mainly concluded from biological effects of islet extracts. It was impossible to demonstrate gastrin within the Brockmann bodies.

In conclusion, there is a fundamental difference in pancreas structure and islet histophysiology between cyclostomes and gnathostomes. Pancreas morphology and islet structure of all higher vertebrates can be derived from a primitive gnathostome type of the pancreas as exemplified by some elasmobranchs.

There is no reason to assume that insulin is absent in any species. A cells and glucagon appear to be absent in the cyclostomes, but A cells could be demonstrated in all species of higher fishes. However, so far glucagon (or glucagonlike activity) has been extracted only from a few teleosts. D cells were demonstrated in all gnathostomes which were studied by appropriate techniques.

Variations in Structure and in Biological and Immunological Activities of the Islet Hormones in Fishes

Insulin

While the structures of the invertebrate "insulins" are still unknown, the available data on their biological and immunological activities indicate that they differ greatly from the vertebrate insulins.

In fishes, we sometimes find an enormous discrepancy between the biological and the immunological activity of insulin. In the light of their evolution this is not surprising, and we also have to bear in mind that even two mammals, the guinea pig and the nutria (both rodents) have insulins structurally very different from each other and from the other heretofore known mammalian insulins.

Cyclostomata

In both biological and immunological activity, the insulin of the hagfish, *Myxine glutinosa*, differs greatly from the insulins of the higher vertebrates. With the double antibody method, the activity is about 0.1% that of bovine insulin. In the fat pad assay with the rat it has an activity of about 7% bovine preparations. Forty mU of anti-ox-insulin serum and 20 mU of anti-codfish-insulin serum from guinea pigs are necessary to neutralize 1 mU of hagfish insulin. According to Falkmer and Matty (1966a) and Falkmer and Wilson (1967), ox and codfish insulin are much less effective in producing hypoglycemia in the hagfish than crude hagfish preparations. In the 4-point mouse diaphragm assay, hagfish insulin shows lower potency than teleost insulins. Studies on the structure revealed that the N-terminal amino acids are glycine (A chain) and probably a basic amino acid (B chain). The A chain contains no phenylalanine; one histidine of the B chain of other known insulins is replaced by a different amino acid. The intact molecule, and especially the A chain, is more basic than heretofore known insulins.

In the lamprey, *Petromyzon planeri*, Ermisch (1965) found strong hypoglycemic activity of the islet extracts of both larval and adult animals. On the other hand, the hemagglutination inhibition test gave evidence of an immunological difference from bovine insulin. High doses of mammalian insulin induce a long-lasting hypoglycemia in *Petromyzon fluviatilis*. In contrast to the situation in mammals, mammalian insulin did not affect muscle glycogen, although it did increase liver glycogen.

Chondrichthyes

In 1922, Macleod showed that dogfish pancreata contain an acid-ethanol extractable substance capable of causing hypoglycemia in rabbits.

Recent studies of Patent (1968) revealed that bovine, dogfish, and raffish (*Hydrolagus*) insulins are effective hypoglycemic agents in both dogfish, *Squalus acanthias*, and ratfish, *Hydrolagus colliei*. Blood glucose levels in dogfish treated with their own insulin returned to normal 3 days following peak hypoglycemia. Dogfish, injected with bovine or ratfish insulins, remained hypoglycemic. Because of the poor condition of the ratfish, it was not possible to determine the duration of the hypoglycemia in this species. Antigenically, shark insulins were found to be different both from ox and codfish insulins in biological and immunological tests; in the passive cutaneous anaphylactic test, the insulin of *Squalus* reacted preferentially with antiserum to chicken insulin.

Actinopterygii

According to Falkmer and Wilson (1967), the insulin of the holostean *Amia* has very low potency in the 4-point mouse diaphragm assay. However, it was readily neutralized by both ox and codfish antiinsulin sera, suggesting that it has an antibody-combining site in common with both of them. The results of the passive cutaneous anaphylactic test are in agreement with this.

The hypoglycemic action of teleost insulins in mammals is well known as is the hypoglycemic effect of mammalian insulin in fishes. However, recent investigations make it clear that there are considerable differences in both biological and immunological properties not only between mammalian and teleost insulins but also among teleost insulins. The blood glucose of teleosts appears markedly resistent to both teleost and bovine preparations. Also, antisera of ox insulin failed to provoke hyperglycemia in *Cottus scorpius*. Even homologous insulin of this species is surprisingly ineffective. On the other hand, Young and Chavin (1967) reported a hypoglycemic effect of low doses of bovine, bonito, tuna, *Hydrolagus*, and *Squalus* insulins in the goldfish, *Carassius auratus*.

Insulin from *Opsanus tau* influenced glucose uptake in rat adipose tissue but not in the muscle. On the other hand, mammalian insulin had no effect on the incorporation of glucose-U-^{14}C into lipids, fatty acids, glycogen, or CO_2 of tissue preparations (heart, liver, and skeletal muscle) of *Opsanus tau*.

Certainly, the differing biological and immunological properties of the insulins result from differences in the molecular structures. A proportion of the primary structure of all insulins studied so far is invariable. However, the common amino acid sequences of the B chains of tetrapod and teleost insulins are only discernible when the whole sequence of the teleost B chain is advanced one position toward the

N-terminal. Nevertheless, phylogenetic conclusions from the biological effects of insulins in the teleosts would be premature because a variety of environmental factors may have a strong influence on the reactions to exogenous insulin.

While the rat is the only mammal known to have two forms of insulin (which only differ in position 29 of the B chain), several teleosts have been shown to produce two different insulins: the bonito, *Katsuwonus pelamis* (?), the flounder, *Pleuronectes flesus*, and the goosefish, *Opsanua tau*. The differences in both biological activity and molecular structure between the insulins of codfish, *Gadus callarias*, from North American and European waters can possibly be explained by racial variations.

In conclusion, the differences in structure and in both biological and immunological properties between the insulin of *Myxine* and gnathostome insulins appear to be greater than the differences among the heretofore known gnathostome insulins. On the other hand, the teleost insulins known so far are rather similar in amino acid composition, but differ markedly from mammalian insulin. Mammalian insulins, as well as insulins from other fish species, are usually hypoglycemic in fishes although very high doses are required. The biological activity of teleost insulins in mammalian test systems shows a considerable species specificity which does not reflect their taxonomic relation. The action spectra of mammalian and teleost insulins appear to be different. No data on nonsuppressible insulinlike activity seem to be available for fishes.

Glucagon

Glucagon was discovered as a hyperglycemic impurity of early preparations of mammalian insulin; yet Geiling and de Lawder (1930) were unable to demonstrate its presence in the insulin preparations of *Gadus morrhua* and *Pollachius virens*. Mosca and associates obtained a hyperglycemic effect of islet extracts from *Scorpaena scrofa* only after inactivation of insulin by 0.1 *N* KOH. Audy and Kerly (1952) found a glycogenolytic effect of islet extracts from *Lophius piscatorius in vitro*; when compared to their yields from mammalian pancreata, the fish glucagon must have been rather inefficient. Weitzel et al. (1953) did not observe any specific glucagon activity (as initial hyperglycemia) in various teleost insulins. Planas and Lluch (1956) obtained a hyperglycemic factor from the Brockmann bodies of *Thunnus thynnus*. Glucagon from *Cottus scorpius* causes hyperglycemia in the fish, but not in the rabbit; and only high doses of ox glucagon evoke hyperglycemia in *Cottus scorpius*. Likewise Wright (1958) observed a rather low sensitivity to the hyperglycemic action of mammalian glucagon in *Lophius piscatorius*. Young and Chavin (1965) found a transient hyperglycemic response to mammalian glucagon in the goldfish, *Carassius auratus*. After injection of mammalian glucagon, slight hyperglycemia was also seen in the elasmobranchs *Raja erinacea* and *Squalus acanthias*, but not in the holocephalian, *Hydrolagus colliei*.

Glucagon from *Cottus scorpius* showed only a faint binding to rabbit antibodies against mammalian glucagon, while the same method failed to detect glucagon in pancreas extracts of *Squalus acanthias* and islet extracts of *Myxine glutinosa*. In conclusion, glucagon appears to have a high biological and immunological species specificity. There do not appear to be any reports on gastrointestinal glucagon in fishes.

Function and Metabolism of the Islet Cells in Fishes

Synthesis, Storage, and Release of Insulin

Studies on the mechanism of protein biosynthesis particularly require tissues of great viability, synthesizing preferably a protein of known structure. Moreover, the tissue must be available in considerable amounts. These requirements are met by the Brockmann bodies. After Lazarow and co-workers had explored this possibility, studies on the islet tissue of teleosts yielded important general information on protein synthesis and particularly on the biosynthesis of insulin.

When islet tissue of teleosts is incubated *in vitro* in the presence of radioactive amino acids or glucose, there is a progressive incorporation of radioactivity into protein. After incubation with radioactive amino acids, Bauer and Lazarow (1961) found that the specific radioactivity of the insulin fraction of the islet tissue of the goosefish, *Lophius piscatorius*, was three times greater than that of the other tissue proteins. Likewise, goosefish islets incorporated radioactivity from glucose-U-^{14}C into many constituent amino acids of the highly purified insulin fraction. Similarly, Hellman and Larsson (1961) observed incorporation of radioactivity from labeled glucose into amino acids of the islet tissue of *Cottus quadricornis*. Corresponding findings on the uptake of glucose carbon and leucine into the insulin fraction of the islets of the toadfish, *Opsanus tau*, were obtained by Humbel and co-workers. Using radioactive isoleucine for *in vitro* studies on toadfish islets, Humbel (1963) also obtained preferential labeling of the A chain of insulin. This would be expected from the absence of this amino acid in the B chain.

The effect of tolbutamide on the incorporation of glucose-U-^{14}C and leucine-^{3}H into the insulin fraction of goosefish islets *in vitro* was studied by Bauer and associates. Contrary to its antidiabetic action in mammals, it inhibited the incorporation of glucose-U-^{14}C, while there was no consistent effect on leucine-^{3}H incorporation.

In extensive studies, Lazarow and co-workers investigated the relationship between insulin synthesis and insulin storage in goosefish islets *in vivo* and *in vitro*. Their findings, especially the results of pulse-chase experiments, suggest the microsomes as primary sites of insulin synthesis, and a subsequent transfer of the hormone to the secretion granules. During these experiments, Bauer et al. (1966) obtained evidence that the secretion granules of their preparations contained not only insulin and glucagon but other proteins as well.

All insulins studied so far consist of two polypeptide chains which are linked by two disulfide bridges. Insulin can be split into its A and B chains and resynthesized; it was even possible to prepare cod and ox "hybrid" insulins by mutual exchange of the isolated chains.

Two different ways have been proposed for the final step of insulin synthesis. Lazarow and co-workers and Humbel (1965) concluded from their own and other authors' findings that insulin is formed from separate chains which are joined together by oxidizing sulfhydryl groups to disulfide bonds. For this process, Lazarow (1965) postulated the existence of a specific enzyme, "insulin zipase." However, recent evidence from studies on the biosynthesis of mammalian insulin favours a different alternative, namely, that insulin is derived from a large, single-chain protein containing intrachain disulfide bonds. Division of this "proinsulin" into two chains is accomplished by cleavage of peptide bonds with resulting loss of the linking fraction of the original single-chain molecule.

The islets of many vertebrates contain very high amounts of zinc. This metal is usually found within the B cells, yet the duck and the rat have a high zinc content in the A and/or D cells. The decrease in zinc concentration during islet cell stimulation suggests a relation between this metal and the islet hormones. Since insulin does not crystallize at physiological pH without heavy metals, zinc may be involved in storage and release of this hormone. On the other hand, glucagon also appears to have a high affinity for zinc.

In teleosts, high concentrations of zinc were found in the endocrine pancreas. Histochemical studies revealed that in *Cottus scorpius* it is located in the islet region which contains B and D cells, and electron microscopically it was demonstrated within the B granules of *Cottus quadricornis*. However, Maske et al. (1956) found no correlations between the concentrations of zinc and insulin within subcellular fractions of the islets of Pleuronectidae, and J. Davidson (1959) found abundant zinc in the degranulated islets of alloxan-treated toadfish. Pihl (1968) suggests that zinc may be implicated in storage and release of insulin rather than in the production of the hormone. He considers the almost complete absence of heavy metal in the islets of the guinea pig, nutria, and *Myxine* as evidence supporting his conclusion. He furthermore points out that the lack of heavy metals in the B cells of these species may result from the absence of certain zinc-affinic histidine residues of the B chains of their insulins. If this conclusion is correct, the insulin of *Petromyzon* must be different from that of *Myxine*.

Metabolic Pathways of the Islet Tissue

As in the liver, glucose freely diffuses into the islet tissue of teleosts and mammals. Thus the possibility exists that insulin release is directly influenced by glycemia and that accumulation of intermediates of glucose metabolism and/or changes in the contribution to these pathways are involved. Both, the Embden-Meyerhof chain and the pentose pathway are found in the islet tissue of both mammals and teleosts. When goosefish islets are incubated *in vitro* in the presence of radioactive glucose, at a "normal" glucose level (25 mg %) the fraction of glucose metabolized via pentose pathway is higher than under hyperglycemic conditions (200 mg %). This eightfold increase of glucose concentration in the medium is followed by a seven-fold increase of the disappearance of glucose from the medium. The estimated amount of glucose oxidized via the Krebs cycle increases fivefold, while little or no increase in the utilization via pentose cycle was observed. Much of the glucose which disappears from the incubation medium may be used for the synthesis of fat and proteins, and some is stored as glycogen.

Possibly, the pentose phosphate cycle is the prevalent pathway in mammalian islet tissue.

Enzymes of the Islet Tissue

The search for glucose metabolites which might be specifically involved in either insulin synthesis or insulin release, stimulated a number of studies on the "enzyme fingerprint" of both mammalian and teleost islet tissue. In teleosts, the activities of hexokinase and phosphatase as well as those of the enzymes specifically involved in glycogen metabolism, pentose shunt, glycolytic (Embden-Meyerhof) pathway, tricarboxylic acid (Krebs) cycle, electron transport system, and transaminations were studied.

Renold et al. (1964) detected significant hexokinase activity in homogenates of the islet tissue of the toadfish, together with 6-phosphogluconic acid dehydrogenase, phosphoglucoisomerase, and phosphomannoisomerase activities.

By microassay, specific glucose 6-phosphatase activity was not found in measurable amounts either in toadfish or in goosefish islets. In toadfish islets, the nonspecific phosphatase activity (pH 6.5) was about half that in liver, kidney, and heart, but slightly higher than in the brain. The alkaline phosphatase activity (pH 9) was about twice the acid phosphatase activity (pH 5). It is noteworthy that most of the histochemical data on glucose 6-phosphatase in mammalian islets confirm the presence of this enzyme. The location of this enzyme primarily in the endoplasmic reticulum of the mammalian B cell led Lazarus and Barden (1965) to the conclusion that it may be involved in insulin synthesis.

The islets of both toadfish and goosefish have a high glycogen content, and glycogen can also be localized histochemically within the B cells. Thus, it is surprising that very low UDPG-glycogen transglucosylase activity was found; however, a comparison with phosphorylase activity is still lacking.

Investigating the enzymes of the pentose shunt in the toadfish, Lazarow and associates found that the islet tissue has the lowest glucose 6-phosphate dehydrogenase activity of all tissues studied except muscle. The 6-phosphogluconate dehydrogenase content of islet tissue was of the same order as glucose 6-phosphate dehydrogenase activity. Islet tissue also had the lowest ratio of glucose 6-phosphate to 6-phosphogluconate when compared with other organs. According to Lazarow et al. (1964b), islet would be the least likely tissue to accumulate 6-phosphogluconate of all the toadfish tissues studied.

Two enzymes of the Embden-Meyerhof pathway were studied in the toadfish. The aldolase activity was about 17% of that found in the muscle, the most active tissue. The 3-phosphoglyceraldehyde dehydrogenase activity was 20% of that of the muscle. The maximal capacity of aldolase was but a small fraction of the maximal capacity of the 3-phosphoglyceraldehyde dehydrogenase.

Of the enzymes of the Krebs cycle, both succinic dehydrogenase and malic dehydrogenase activities were studied in toadfish islets. The activities of both enzymes were rather low when compared with other tissue. Maske et al. (1956) found insulin and succinic dehydrogenase in parallel concentrations within the subcellular fractions of islet tissue of Pleuronectidae.

Lindall (1962) determined the concentrations of both reduced and oxidized forms of diphosphopyridine and triphosphopyridine nucleotide of toadfish islet tissue. The ratio of the oxidized-to-reduced form was 0.40.

For the electron transport system of the toadfish islets, Lazarow and co-workers obtained evidence that cytochrome c may exist predominantly in the oxidized state. This was suggested by the observation that cytochrome oxidase activity was rather low when compared with the activities of the other dehydrogenase systems which reduce cytochrome c. These findings, together with the high content of oxidized nicotinamide-adenine dinucleotide suggest that the redox potential of the islet tissue is high. According to Lazarow (1954, 1965) and Lazarow et al. (1964b), this would be favourable for the oxidation of sulfhydryl precursors to their disulfide form during insulin synthesis. Furthermore, the high ratio of NAD^+ to NADH would make it less likely that the islet tissue inactivates alloxan by reduction to dialuric acid.

In general, the comparison with other tissues indicates that the relative activities of the enzymes of the glycolytic, tricarboxylic, and electron transport pathways are low, despite the high capacity of the islets to utilize glucose. As pointed out by Lazarow (1965) and Lazarow et al. (1964b), these enzymic activities were measured under optimal conditions *in vitro*; thus, the findings presented may not necessarily reflect the relative roles of these enzymes in the intact cell.

As an intensive conversion of glucose into amino acids was found, Hellman and Larsson (1962) studied the activities of glutamic-oxalacetic transaminase, glutamic-pyruvic transaminase, and in addition ornithine carbamyltransferase in the islet tissue of *Cottus quadricornis*. When compared with the exocrine pancreas and liver, the islet tissue showed relatively high activities of these enzymes. The high level of ornithine carbamyltransferase in endocrine and exocrine pancreas is a striking contrast to what would be expected from findings in mammals where the activity of the latter enzymes in the liver was found to be 2000 times greater than in the pancreas.

It should be noted that the data presented here are not specific for the B cells but derived from the study of the Brockmann bodies which, of course, also contain other epithelian elements.

Physiological Role of the Islets in Fishes

Islet Changes under Normal Conditions

Islet changes with age

Apparently, a study on the embryonic development of the endocrine pancreas of fishes with modern methods is lacking. The literature on the development of the pancreas of the cyclostomes is well covered by Ermisch (1966). During the metamorphosis from the larval to the adult form in *Petromyzon planeri* there is a conspicuous change in the islets. The islets of the larva contain only B cells and a few agranular elements; during the metamorphosis a second type of granular cell appears in large numbers. As pointed out by Ermisch (1966, 1967), this may well be associated with the end of food uptake of the adult animals and the subsequent utilization of lipid reserves.

Senile, castrated Pacific salmon, *Oncorhynchus nerka kennerlyi*, which had survived 1 to almost 4 years beyond the normal life span showed marked hyperplasia and hypertrophia of the islets.

Seasonal islet variations

Pallot and associates report very striking alterations of the exocrine and endocrine pancreas of several teleosts. In the carp, *Cyprinus carpio*, they describe a change from a prevalent B-cell activity in summer to an A-cell hypertrophy in winter. This was not seen in Italian or in Japanese specimens of the same species. Dancasiu (1960) finds islet neoformation in Roumanian carps during May and June; yet this takes place in a way quite different from the islet alterations described by Pallot and associates. Ghittino (1961) reports seasonal changes in the relative proportions of "A" and "B" cells in the islets of another cyprinide, *Leuciscus suffia muticellus*. Honma and Tamura (1968) observe no seasonal changes in the islet composition of *Salvelinus leucomaenis pluvius*, but an increase in the number of small islets after the breeding season. In a sedentary population of *Salmo trutta*, Schneider and Epple (1969) did not observe clear signs of a seasonal cycle. Also, Falkmer (1961) did not see seasonal variations in the marine teleost *Cottus scorpius*. Maske et al. (1956) obtained higher yields of insulin and zinc from the Brockmann bodies of Pleuronectidae in December than in April; they explain this with the change in nutritional conditions.

It appears difficult to interpret these differing findings. Possibly, the interaction of local factors (such as temperature and food supply) with the life cycle of a species plays a role. In some cases, increased food intake and growth during summer may lead to a high seasonal requirement of insulin; in other cases, when animals are active also in winter (e.g., *Salmo trutta*), the insulin requirement may be similar at all seasons.

Islet changes and migration

Islet hyperplasia occurs both in Pacific salmon, *Oncorhynchus nerka*, and in the rainbow trout, *Salmo gairdneri*, during the anadromous spawning migration, but it is also found in catadromous eel, *Anguilla japonica*, from the depth of the Japan Sea. Nevertheless, islet hyperthrophy occurs in spawning nonmigratory rainbow trout.

Experimental Islet Studies

Hyperphagia and starvation

According to Hess (1935), reduction of the number of islets can be induced in the rainbow trout, *Salmo irideus*, by (1) addition of fat or fat and carbohydrate to the diet, (2) overeating, and (3) lack of muscular activity. "The greatest reduction of islets was observed after addition of fat and carbohydrate to the diet; overeating led to a greater reduction of islets than the lack of exercise. Higher islet counts were obtained by feeding beef liver than with either pig spleen or beef heart. Baron (1934) finds no changes in the islets of starving sticklebacks, *Gasterosteus aculeatus*. Starvation causes a reduction of islet size in the eel, *Anguilla anguilla*.

Glucose administration

Glucose tolerance

Whereas Young and Chavin (1965) report a short mammalianlike response to glucose injections in goldfish, *Carassius auratus*, kept at a rather high temperature (25°C), all other investigators observed a rather slow return to normal levels after glucose injections in fishes. A slow response seems to be characteristic of poikilotherms in general. In cyclostomes (*Petromyzon fluviatilis* and *Myxine glutinosa*), only small amounts of urinary glucose are found after glucose loading, despite a sharp drop of the blood sugar after several hours; Bentley and Follet (1965) and Falkmer and Matty (1966a) conclude from this observation that the animals must have mechanisms concerned with the control of blood glucose concentration. However, previous isletectomy did not influence the glucose loading curve in *Myxine*. In several elasmobranchs, in the holocephalian, *Hydrolagus colliei*, and in the teleosts, *Cottus scorpius* and *Clarias batrachus* glucose levels did not return to normal before a period of 9 hr to several days.

Effects of a glucose-containing medium

Kohler (1963) kept goldfish in 2% glucose or 2% fructose solutions; he observed fatty liver only in animals kept in glucose. Addition of insulin to the medium prevented fatty liver, and adequate controls suggest that this effect was a specific action of the hormone. Also, in a 0.2.% glucose solution, goldfish developed hyperglycemia together with hepatic steatosis and an increase of the sugar level of the muscles. High glucose concentrations also stimulate insulin release from toadfish islet tissue *in vitro*.

Islet changes after glucose injections

In lampreys, Barrington (1942) and Ermisch (1966, 1967) observed transient glycogen infiltration and vacuolization followed by necroses after glucose injections; the intrafollicular, colloid-containing lumina of the larval *Petromyzon planeri* disappear, but the gomori-negative granular islet cells seem practically unaffected. In *Myxine*, a long-term injection of glucose led to B-cell degranulation but not to the disappearance of intrafollicular colloids. No effect of glucose injection was observed in the pancreatic islets of *Squalus acanthias* and *Hydrolagus colliei*. In the teleost, *Clarias batrachus*, glucose injection caused temporary degranulation and "fusion" of the B cells.

Isletectomy and pancreatectomy

Destruction of the islet tissue of the larval *Petromyzon marinus unicolor* by cautery caused hyperglycemia, while surgical removal of the islet organ of *Myxine glutinosa* had no effect on the blood sugar level. It remains to be seen if the different reactions of *Petromyzon* and *Myxine* result from scattered B-cell-like elements in the bile duct and possibly also in the gut of the latter species. Pancreatectomy in elasmobranchs causes hyperglycemia, which can be alleviated by hypophysectomy. Removal of the compact pancreas of the eel was followed by inconstant glucosuria. In teleosts removal of the Brockmann bodies leads to hyperglycemia, glucosuria, and increase in liver lipids. However, in *Cottus scorpius* the occurrence of hyperglycemia is more evident in fed than in starved animals; insulin alleviates hyperglycemia, while hypophysectomy shows less consistent results.

Effects of islet cytotoxins

Alloxan

Ever since the observation that alloxan selectively destroys the pancreatic B cells in the rabbit, this compound has been widely used to induce experimental diabetes or to support the histological identification of the B cells. There is even a report of alloxan "diabetes" in a clam. Representatives of both groups of the cyclostomes have been treated with alloxan. In lampreys (*Petromyzon planeri* and *Petromyzon fluviatilis*), alloxan injection sometimes causes B-cell destruction; however, very high doses are required, and other organs (liver, kidney, and intestines) may also be damaged. In *Myxine glutinosa*, extremely high doses of alloxan led only to hyperglycemia and damage of B-cell groups in a relatively small number of the experimental animals. There was no correlation between B-cell destruction and hyperglycemia. These observations might result from special circulatory conditions, e.g., imperfect mixing of blood and slow circulation. All animals, regardless of the state of islet damage or glycemia, showed a very low hematocrit.

The reports on B-cell destruction by alloxan in the chondrichthyans are rather controversial. After intraperitoneal injections, Saviano (1946, 1947a) finds an increase of blood sugar but no islet damage in sharks. In highly hyperglycemic animals there was a remarkable development of the A cells and a clear increase of the cytoplasmatic granules. Paradoxically, the exocrine pancreas was damaged by the drug. Even with high intramuscular doses of alloxan. Kern (1966) was unable to produce any alterations in the blood sugar level or the islet organ in *Scyliorhinus canicula*, *Raja asterias*, and *Torpedo marmorata*. However, he also observed lesions in the exocrine pancreas, as well as in the interrenal organ, adrenal bodies, and kidney tubules. Clausen (1953) reports B-cell destruction in two specimen of *Scyliorhinus canicula* after subcutaneous injection of relatively low doses of alloxan; the A cells of these animals appeared hypertrophied as in sharks, as reported by Saviano (1947a). In *Hydrolagus colliei*, Patent (1968) describes selective damage of the B cells after injection of only 300 mg/kg body weight into the cannulated dorsal aorta.

It appears difficult to explain the differences between these findings. Perhaps the manner of injection is of paramount importance in chondrichthyans since alloxan may be immediately inactivated by the abundance of urea, forming alluranic acid. However, the urea content in the blood of the holocephalians is at least as high as that of the selachians. Thus, the results of Patent (1968) show that urea is not responsible for the difference.

Many studies deal with the diabetogenic action of alloxan in teleosts. Several investigators describe B-cell lesions following alloxan injection, the degree of these lesions varying greatly. For example,

Doerr (1950) concludes that a specific alloxan diabetes cannot be produced in *Cyprinidae*; on the other hand, Falkmer (1961) finds hyperglycemia and B-cell destruction for as long as 3 weeks after alloxan treatment in *Cottus scorpius*. The morphology of B-cell destruction in teleosts is well documented in the publications of Murrel and Nace (1959) and Falkmer (1961). Matty and Qureshi (1967) observed in alloxan-treated islets of *Cottus* a decline of the RNA content. Since several investigators found destruction of other tissues such as liver, kidney, and exocrine pancreas, suitable histological controls are necessary to determine whether the hyperglycemia, after alloxan injection, results from selective B-cell destruction or the toxic effects on other organs. Also, the question arises as to whether the term "alloxan diabetes" is justified. Hyperglycemia without concomitant impairment of the lipid metabolism would not correspond to the diabetic syndrome in mammals. La Grutta (1950), however, reported hepatic steatosis in alloxan-treated specimens of *Scorpaena scrofa*. Since this could be prevented by insulin application, it suggests that the alloxan diabetes in teleosts is similar to that in mammals, despite some differences in the biological action of mammalian and fish insulins.

These conflicting results and varying interpretations do not permit a general statement on the action of alloxan in teleosts. Species differences in the sensitivity of the B cells to alloxan are well known in mammals and may also play a role in teleosts. Furthermore, the method of blood sugar determination, as well as the fishes' response to handling, the manner of application of alloxan, its quality and instability above pH 6, and the water temperature are factors which must be considered in interpreting the available data. However, in some teleosts it appears that alloxan diabetes can be elicited.

There are several theories on the mechanism of the B-cell destruction by alloxan. Lazarow's "sulfhydryl theory" (1954) stimulated a series of investigations on the mechanism of alloxan action; both *in vivo* and *in vitro* studies have been reported for teleosts. Lazarow and associates conclude from recent studies on fish islets that alloxan acts selectively on the cell membrane of the B cells. The permeability of the cell membrane may be damaged at a site involved in sugar transport.

Other cytotoxins

Several drugs have been reported to selectively destroy the A cells in mammals and/or birds. Four of these drugs (cobaltous chloride, Synthalin A, *p*-aminobensolsulfonamide isopropyl-thiodiazol (IPTD), and sodium diethyldithiocarbamate) were tested in fishes. Ermisch (1966) did not observe a specific effect of $CoCl_2$ on the islet tissue of lampreys (*Petromyzon planeri* and *Petromyzon fluviatilis*). Likewise, Schirner (1963b) and Falkmer and Winbladh (1964b) failed to achieve islet damage by application of $CoCl_2$ in *Myxine*. Schirner (1963b) reports "A"-cell lesions in *Myxine* after injection of Synthalin A; these were not seen in the experiments of Falkmer and Winbladh (1964b). Fodden (1956) did not obtain any islet changes in the toadfish after $CoCl_2$ injections. However, in *Scorpaena scrofa* injections of $CoCl_2$ and sodium diethyldithiocarbamate led to a decrease in the extractable, hyperglycemic activity of the Brockmann bodies. *Gambusia holbrooki* has a third type of islet cell (D cell?) which is especially sensitive to $CoCl_2$. In his earlier investigations, Falkmer (1961) did not observe a cytotoxic effect of $CoCl_2$, Synthalin A, or IPTD on the islets of *Cottus scorpius*. However, in subsequent studies, Falkmer et al. (1964a,b) found that cobalt is selectively concentrated in the central islet region of this species. Repeated injections of $CoCl_2$ destroyed the B cells and perhaps also the D cells, while A and agranular (C) cells were unaffected. Alterations in the A cells after application of $CoCl_2$ may be secondary effect to changes in the blood sugar.

Block et al. (1964) used the toadfish to analyze glycemic changes induced by methylglyoxal bis(guanylhydrazone), a chemotherapeutic agent against human leukemia and solid tumors which also causes profound hypoglycemia. Low doses caused hypoglycemia; high doses caused hyperglycemia and A-cell degranulation. The authors suggest that the drug induces hepatic glycogenolysis and increases glucose uptake in the peripheral tissues of the toadfish.

Effects of hormone injections and hypophysectomy on islet tissue

In lampreys (*Petromyzon planeri* and *Petromyzon fluviatilis*), insulin injections sometimes led to B-cell stimulation (degranulation and nuclear enlargement) and sometimes to B-cell atrophy. There was no clear effect on the second type of granular islet cell. In *Myxine glutinosa*, injections of thyrotropin, corticotropin, prolactin, thyroxine, adrenalin, glucagon, or insulin evoked no histological changes in the islet tissue; also, hypophysectomy did not affect the islet tissue. Butler (1940) reports an enlargement of the islets in the goldfish, *Carassius auratus*, after injection with a pituitary brei of the same species. No cytological changes could be observed in the pancreatic islets of *Fundulus heteroclitus* following removal of the pituitary or after treatment with mammalian growth hormone. In *Cottus scorpius*, injections of adrenalin, porcine growth hormone, glucagon, and insulin had no effect on the islet histology, while hydrocortisone led to occasional B-cell damage. Hypophysectomy did not influence the islet histology in this species. Khanna and Mehrotra (1969b) observed shrinkage of B cells in *Clarias batrachus* after insulin injection. In the eel, *Anguilla anguilla*, thyroxine injections decreased the number of islets.

Effects of Exogenous Islet Hormones in Fishes

Insulin

Blood sugar

Most authors have observed that exogenous insulin has a hypoglycemic action in fishes and that excessive doses of insulin lead to hypoglycemic convulsions and/or death. Insulin hypoglycemia was provoked in cyclostomes, in chondrichthyans, and actinopterygians. The data on duration and degree of hypoglycemia and on the onset of hypoglycemic convulsions vary greatly. Most investigators observed a slow response, the peak of hypoglycemia or convulsions occurring 1 or 2 days after injection, or even later. This agrees with the observations on insulin action in other poikilotherms. Seshadri (1967) reported a mammalianlike hypoglycemic response in *Ophicephalus striatus* with return to normal blood sugar levels within 3 hr. Falkmer and Wilson (1967) observed a weak hypoglycemic response even to species specific insulin in *Cottus scorpius*; Tashima and Cahill (1964) failed to obtain any hypoglycemia in *Opsanus tau* at a very high dosage of bovine insulin. Young and Chavin (1967) observed a moderate hyperglycemia after 1000 U/kg and 5000 U/kg bonito insulin in the goldfish, while 1 U/kg produced significant hypoglycemia.

Glycogen

In lampreys, mammalian insulin leads to an increase of ambiguous behaviour of liver glycogen; it causes little or no increase in muscle glycogen. In *Petromyzon*, exogenous insulin causes a decrease of the high glycogen content of the brain. In *Lampetra fluviatilis* injection of insulin decreased the glucose 6-phosphatase activity in the liver. In chondrichthyans, exogenous insulins have no effect on either liver or muscle glycogen or they lead to an increase of glycogen in one or both organs. Insulin also increases the glucose uptake of dogfish branchial muscle *in vitro*. In teleosts, the effects of exogenous insulin on the glycogen content of liver and muscle vary greatly. Root et al. (1931) found in *Stenotomus chrysops* a transient increase in liver glycogen, followed by hypoglycemia and an increased glycogen deposition in muscles. Tashima and Cahill (1964) found no effect on glucose incorporation into toadfish tissues *in vitro*. Seshadri (1967) observed in *Ophicephalus striatus* a peak of both liver and muscle glycogen deposition 90 min after injection, and returning to normal after 3 hr, while the blood sugar level showed an inverse response. Increased dosage of insulin caused an increase in both liver and muscle glycogen. In *Clarias lazera*, Yanni (1964) found a slight increase in muscle glycogen and a slight decrease in liver glycogen, while insulin plus glucose increased both muscle and liver glycogen. However, double dosage of insulin with glucose had a similar effect to insulin alone, i.e., a strong increase in muscle glycogen and a slight decrease in liver glycogen. Glucose administration without insulin caused the largest increase in both muscle and liver glycogen. The glycogen content of several other tissues behaved like the muscle glycogen under these experimental conditions. As in the lamprey, the high glycogen content of the brain of *Scorpaena* decreased after insulin injections. Thus, insulin may cause either increase or decrease of liver glycogen, while muscle glycogen is either unaffected or increased. In fishes with great glycogen storage in the brain the latter is depleted.

Lipids

Apparently, there are no data available on the influence of the islet hormones on the lipids in cyclostomes and elasmobranchs, although lipids seemingly play an extremely important role in these fishes. In teleosts, isletectomy, B cell destruction by alloxan, or exposure to a glucose-containing medium increased the liver lipids. This could be prevented by application of insulin. Insulin also decreased both liver and muscular lipids of both normal and glucose-injected specimens of *Clarias lazera*. In heart, liver, and muscle preparations of *Opsanus tau* there was no effect of mammalian insulin on the *in vitro* incorporation of glucose into total lipids or fatty acids.

Proteins

In *Ophicephalus striatus*, insulin injections result in a decrease of muscle free amino acid and an increase of protein-bound amino acids.

The above observations make it rather discouraging to draw conclusions on the physiological role of insulin in fishes. There are many factors which possibly influenced the results of the investigations such as species specificity of the hormone, purity of the hormone (glucagon content), method of hormone application, dosage of the hormone, duration of the experiment, intervals between determinations, method of glucose or glycogen determinations, handling and housing of the animals, nutritional state of the animals, metabolic rate of the species, sex of the animals, season, water temperature, and various combinations thereof. With this in mind, it becomes clear that many more data are necessary to evaluate the physiological role of insulin. In future research, the use of species specific hormones as well as the measurement of endogenous insulin under various natural and experimental conditions may become indispensable.

Glucagon

In *Lampetra fluviatilis*, glucagon injections had no initial effect, but after a delay of 4 hr decreased the blood glucose level. In *Myxine glutinosa*, glucagon had no effect at all. In chondrichthyans, high doses of glucagon may produce a slight hyperglycemia. Teleosts respond to glucagon with hyperglycemia. However, since species specific glucagon was much more effective than ox glucagon in *Cottus scorpius*, the specificity of the hyperglycemic effect of mammalian glucagon in fishes is difficult to assess. Hyperglycemia is easily evoked by many factors. When given to toadfish tissues *in vitro*, glucagon caused no changes in glucose metabolism in heart or skeletal muscle. In the liver, it (1) depressed conversion of glucose into total lipids and CO_2 (2) increased conversion of glucose into glycogen, (3) stimulated liver glycogenolysis at the same time, and (4) from acetate or alanine it stimulated gluconeogenesis.

As in the case of insulin, the available experimental data do not yet permit to draw conclusions on the physiological role of glucagon in fishes.

11

ECOLOGICAL ADAPTATION

Environmental conditions are characteristically variable. A relative constancy is found in only a few habitats, such as those of some parasites or those of animals living in the depths of the ocean. Over the earth's surface and in the upper reaches of most aquatic environments animals experience a rhythm of changing diurnal and seasonal conditions. Successful living demands a continual adjustment or regulation of internal media, metabolism, and activity in relation to these cycles of environmental change. The diurnal variations may be of considerable magnitude, and animals must be able to tolerate a wide range of temperature, moisture, or other environmental conditions if they are to survive. The seasonal variations—at least in the temperate and frigid regions of the world—are even greater.

Organisms must gradually acclimatize and adjust their physiological machinery to operate at some seasons under conditions which would be quickly lethal at others. Such environmental extremes may be great enough to impose definite restrictions on the animal's metabolism and activity. Temperature may be so low and the available food so meagre that it is impossible for the small bird or mammal to maintain its homoiothermic condition. Hibernation or migration provides an escape from such an extreme. Water conditions may be such that some fish and amphibians can only find suitable places and conditions for reproduction during a relatively brief period each year, so that breeding must be cyclical. Food of the right quality and quantity may be only seasonally available for rapid growth and development. The success of many species depends on precisely timed physiological cycles which prepare the animal for reproduction, growth, migration or hibernation at the appropriate season.

The physiological interrelationships are complex and the endocrine system is intimately involved in their control. The evolution of an organism capable of adjusting to a characteristically variable environment and of exploiting its most favourable aspects has involved, in a very real way, the endocrine system. The coordination of these physiological rhythms is a chemical coordination dependent on cyclical changes in both endocrine activity and in sensitivity of effector organs. The basic rhythm of glandular changes is frequently regulated by the environment although, even under relatively constant environmental conditions, cyclical changes may sometimes occur.

Seasonal Cycles in Endocrine Activity

Reproduction of fish is always cyclical in the temperate and arctic regions, and frequently so in the tropics. Changes in the endocrine as well as gametogenetic tissue of the gonads are evident, and the expected cycle in the pituitary which produces the gonadotropic hormones has been often described. In addition, coordinated changes in thyroid and interrenal are known. No attempt will be made here to review this literature since several detailed summaries are now available, and it seems probable that similar morphological changes will be found wherever a careful search is made. Of more particular interest to the present review are the controlling environmental mechanisms.

In the temperate and frigid zones the most constant seasonal variable is the changing length of the daily photoperiod. In many cases it has now been shown that the basic endocrine rhythm is wholly or partially dependent upon radiant energy. Physiological changes associated with endocrine activity have now been induced by control of the photoperiod in cyprinids, salmonids, poeciilids, centrarchids, sticklebacks, and probably other species of fish. Seasonal temperature changes are almost as regular as light changes in the temperate and frigid regions of the world and are usually associated with light in the control of the endocrine rhythms. The literature is discussed in several of the papers quoted above.

In the tropics, cycles of temperature and illumination are less marked and, near the equator, are probably too weak to form useful or reliable triggers for endocrine change. Nevertheless, variable seasonal conditions—often associated with periodic flooding of rivers—may produce much more favourable conditions for reproduction and feeding at certain times of the year. Seasonal breeding and an associated endocrine cycle may be found in the tropics as well as in other parts of the world. The controlling mechanisms have been sought by several workers and, although not fully elucidated, sharply rising waters associated with the rainy season seem to provide the necessary cues—perhaps both physiological and ethological—for some species.

INFLUENCE OF HORMONES ON ENVIRONMENTAL TOLERANCE AND RESISTANCE

Temperature

Temperature is more likely than any other environmental variable to change a tolerable environment suddenly into a lethal one. Although the aquatic habitat is better buffered than the terrestrial one against sudden temperature changes; nevertheless, many organisms, including fishes, may be killed by unseasonal and sudden changes in temperature. Any mechanism which enables a fish to resist extremes of temperature for short or for longer periods will have high survival value. Gradual acclimatization to higher or lower temperatures associated with the seasonal cycles is an important phenomenon but may be inadequate to protect a fish during an unusually cold or warm period. Increased resistance for even a brief period may mean the difference between life and death.

Seasonal variations in the temperature resistance of goldfish

Seasonal variations in the temperature resistance of goldfish were noted during the course of an investigation of dietary factors involved in the temperature relations. These variations were independent of dietary or temperature history. Goldfish maintained for long periods at 20°C on standard diets were found to be more resistant to a sudden exposure to low temperatures (2°C) during the winter and relatively more resistant to sudden exposures to high temperatures (36°C) during the summer. It seemed reasonable to assume that this might be another example of a photoperiodically controlled physiological mechanism, and that in some way (probably through the pituitary gland) the changing lengths of day and night were promoting the development of a factor or factors responsible for the variable temperature resistance.

This hypothesis was tested experimentally by comparing the temperature resistance of groups of goldfish held for varying periods under different photoperiods. Fish maintained for one month or longer under daily photoperiods of 16-hour illumination (about 60 foot candles) alternating with 8 hours of darkness were relatively less resistant to cold than fish maintained under the reverse conditions of illumination. The "long day" fish behaved like summer fish and the "short day" fish like winter fish when exposed to sudden changes in temperature.

The initial tests were carried out in October and November. Subsequently it was found that the differences in temperature resistance produced by photoperiod control are less marked and, in some cases, are not evident during the spring and summer breeding season. An analysis, however, based on a series of experiments extending over more than a year confirms the theory that the seasonal variations in temperature resistance observed earlier are real and that they are most likely controlled by the seasonal cycle of illumination. Factors responsible for the differences in temperature resistance were next investigated.

Thyroid hormone and temperature resistance of fish

The stimulating effect of a cold environment on the thyroid metabolism of a mammal is well known and related to the recognized calorigenic action of thyroid hormone in this group of animals. Investigations on fish have produced conflicting results, both with respect to the metabolic and the temperature effects of this hormone. The metabolic effects need not concern us here.

Olivereau (1955a, b, c) examined thyroids from several species of fish maintained at higher (about 20°C) and at lower temperatures (5° to 10°C) and found no particular differences in the histological pictures of carp (*Cyprinus carpio*), tench (*Tinca tinca*), eel (*Anguilla anguilla*), chub (*Mugil auratus*), or dogfish (*Scyllium canicula*) maintained at these two levels of temperature for periods of several days up to two months during different seasons of the year. In the case of the trout (*Salmo gairdneri*), however, the glands seemed slightly more active at the lower temperatures after 9 to 10 days. These observations were fully confirmed in radioautographic studies with I^{131}. The trout held at the lower temperatures showed a slightly greater ability to trap this isotope.

Fortune (1955) came to a diametrically opposite conclusion in her comparisons of thyroid function in tropical (*Lebistes*) and temperate (*Phoxinus*) fishes. Her findings are, however, rather variable. In the case of *Phoxinus* little or no difference in thyroid activity was evident during the first 10 days of temperature control, whereas in the case of *Lebistes* the differences disappeared after three weeks of controlled temperature. Barrington and Matty (1954) likewise found an increase in the height of the thyroid follicular epithelia of *Phoxinus* maintained at higher temperatures. Epithelial cell height was measured in groups of fish maintained for 57 days in darkness at 3, 14, and 26°C. Differences were statistically significant and indicate that, under the conditions of these experiments, thyroid activity was greater at higher temperatures. Berg, Gorbman, and Kobayashi compared the uptake of radioiodine by thyroid glands of *Fundulus diaphanus* and *Umbra limi* at different temperatures. Reduction of aquarium temperature decreased thyroidal I^{131} uptake of *Fundulus* but increased that of *Umbra*.

It should be remembered that the thyroid is an extremely labile organ and its histological picture may be expected to change in relation to many different factors. Indeed, thyroid activity in fish has been associated with seasons of rapid growth, sexual development, osmotic changes in the environment and with transformations such as the smolt change in salmonids and the metamorphosis of flatfish. Temperature may perhaps be added to the list of factors which modify thyroid activity, but experiments designed to prove this must be carefully controlled to exclude these other variables. For this reason the many studies of seasonal changes in thyroid activity are not considered here.

Thus, the evidence concerning the effects of temperature on thyroid activity of fish is conflicting. The same is true of poikilotherms in general. The literature on the reptiles and amphibia has been summarized. In general, it is indicated that the thyroid-temperature effect is slight. In some species, however, the thyroid gland is definitely stimulated by a rise in temperature while in others stimulation is associated with a fall in temperature. The adaptive significance of these differences is not yet clear.

Thyroid activity in goldfish maintained under controlled photoperiods

Thyroid activity was measured by injecting tracer doses (2 to 4 microcuries) of radioiodine intraperitoneally and determining at intervals the amount of I^{131} accumulated in the thyroid region. The data are reported more fully elsewhere. This technique involves a wet ashing of samples containing variable amounts of muscle and the other tissues associated with thyroid. The activity subsequently determined with a Geiger counter gives extremely variable results due partially to self-absorption of beta rays. In addition, a part of the thyroid of the goldfish may be located in the head kidney and this portion is not sampled in the present technique. These two experimental errors may be expected to produce variable results.

Although there is no statistically significant difference between the two groups, the fish which experienced short day periods consistently show a slightly greater uptake of I^{131}. These are the goldfish which best resist low temperatures. It might be argued that the results suggest greater thyroid activity associated with periods of greater low temperature-resistance, and that these findings are in accord with those

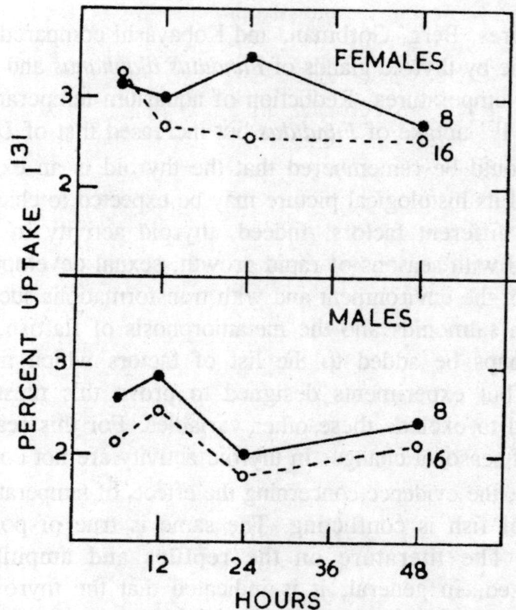

Fig. 11.1. Percent uptake of I^{131} by the throat thyroid gland tissues of goldfish maintained under daily photoperiods of 8 hours (solid line) and 16 hours (broken lines).

of Olivereau (1955a, b, c) for the trout and Gorbman et al. for *Umbra*. Contrary to this argument, Fortune (1956) failed to find a temperature effect on the histology of the goldfish thyroid. Additional properly controlled experiments obviously are needed.

Thyroid treatment and temperature resistance in goldfish

Goldfish have been immersed for varying periods in synthetic sodium thyroxine (B.D.H., 1:2,500,000), 0.05% thiourea, 0.05% thiouracil or injected intraperitoneally with thyrotropic hormone (0.6 U.S.P. units per fish). In one series of experiments the fish were kept under ordinary laboratory conditions of illumination, whereas in a second series the photoperiods were controlled as described in the temperature resistance tests.

Natural light conditions

Fish were tested after 3, 7, 14, and 21 days of treatment. In these tests, samples of 17 to 20 fish were taken from their aquaria at 20°C and placed immediately in a cold bath at about 2°C. This bath was gradually cooled over a period of about 1 hour to 1°C and maintained at this temperature. Different groups were distinctively marked prior to the test, and all comparisons were of fish tested in the same tank at the same time. Fish were removed from the bath when they no longer responded to a standard mechanical stimulation. They were placed directly in an aquarium at 20°C, where they sometimes recovered (narcotized fish). Consequently three groups of fish may be distinguished in this test: "removals," "narcotized," and "truly dead" fish. The values are the lengths of time after immersion in the cold bath before 50 per cent of the fish were removed. The values in brackets are ratios for the survival times of "treated" in relation to "control" fish. The variations in absolute survival times are, at least in part, due to small differences in the temperature of the lethal bath at the different times of testing. However, it is again emphasized that all comparisons are based on fish subjected to identical conditions since they were always tested together in the same tank at the same time.

Table 11.1. Removal time in hours for 50% of samples in chill bath

	Control	*Thyroxine*	*Thiourea*	*Mixed thyroxine/thiourea*
Summer fish				
3 days	14.8	26.5 (1.56)	31.0 (2.09)	24.2 (1.64)
7 days	21.3	22.3 (1.05)	35.0 (1.64)	22.3 (1.05)
14 days	11.2	16.8 (1.50)	21.2 (1.89)	11.9 (1.06)
21 days	1.9	2.5 (1.32)	10.7 (5.63)	3.7 (1.95)
Winter fish				
3 days	9	5 (0.55)	4.5 (0.5)	3.5 (0.39)
7 days	12	7.5 (0.65)	10.0 (0.8)	6.5 (0.54)
14 days	11	6.5 (0.6)	10.0 (0.9)	7 (0.63)
21 days	14	16.0 (1.14)	11.0 (0.8)	7.5 (0.53)

It is evident from table 11.1 that in four successive tests carried out in July, all three treatments exerted a marked protective action. In comparable tests in November, however, the control fish almost always resisted the low temperature longer than the experimentals.

Two contradictions are apparent in these data. The effects are entirely opposite at the two seasons; and there is an apparent similarity in action of thyroxine and the anti-thyroid drug, thiourea.

These data were not immediately accepted because of the contradictory nature of the results. However, the findings for the photoperiod fish, presented in the next section, are consistent with these and indicate that the observed results are valid—whatever may be their explanation.

Controlled photoperiods

In this series, thyrotropic hormone and thiouracil were also included. The long-day fish, like natural summer fish with low cold resistance, showed a marked increase in cold resistance when treated with either thyroxine or thiourea. Thyrotropic hormone likewise increased the cold resistance. This is fully in accord with the theory that thyroid hormone favours cold resistance. The thiouracil effect was also consistent with this theory and with the recognized anti-thyroid effect of this compound. The thiourea effect on the long-day fish—as in the case of the summer fish—is puzzling and suggests that this material is exerting some pharmacological effect unrelated to the thyroid.

The effects of these treatments on the short-day fish are less consistent but in agreement with those for the long-day fish. Again, it should be recalled that low temperature resistance is naturally high in these 8-hour fish—perhaps maximally high in many of the fish. In general, however, thyroxine and TSH increased the survival time of the short-day fish, at least in the latter part of the test (75 per cent

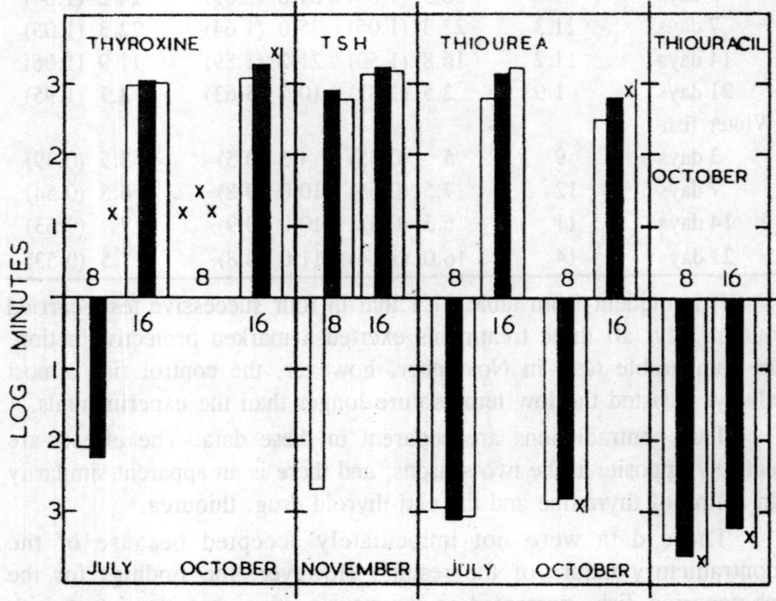

Fig. 11.2. Effects of thyroid treatment on the cold resistance of goldfish.

removal time), while both anti-thyroid materials decreased the cold resistance. Although results from the 8-hour fish are less consistent than those for the 16-hour fish, they too support the theory that thyroid hormone favours cold resistance and that the seasonal changes in cold tolerance are related, at least in part, to thyroid activity.

Heat resistance experiments comparable to those just reported for cold resistance have not yet been completed. Several workers have, however, investigated effects of thiourea on the high temperature resistance. Increased resistance, following treatment with thiourea, has been reported in cold-adapted but not warm-adapted goldfish, in *Phoxinus* and *Lebistes* and in *Coregonus* larvae. Auerbach (1957), on the other hand, recorded decreased heat resistance in thiourea treated *Lebistes*, *Platypoecilus* and *Xiphophorus*. One should hesitate, however, to attribute these differences entirely to the anti-thyroid action of thiourea in view of the data just presented for cold resistance.

Salinity

The mineral content of an aquatic environment is not subject to sudden local fluctuations of the lethal magnitude just described for temperature. Many species of fish, however, are adapted to live in waters differing greatly in their mineral content. In some cases, the same individuals move freely back and forth between fresh and saline waters and exploit the resources of both environments. Several hormones are evidently involved in the essential physiological adaptations.

Two different kinds of physiological problems may be encountered by species of fish which exploit waters of different mineral content. The first of these is the osmotic problem. Species living both in fresh and sea water must be able to excrete copious amounts of osmotic water and absorb salts when in the hypotonic medium, but they must also be able to drink sea water, reduce urine output, and excrete salt while in the hypertonic marine environment. The second problem is one of acquiring essential minerals from waters in areas where there is relatively little dissolved solid. Iodine, for example, is essential for the production of thyroid hormone. Thyroid function is sometimes seriously impaired in fishes living in goitrogenic areas of the world. It has been suggested that the iodine content of the waters may limit growth and reproduction of large, populations of certain fish species.

The osmoregulatory functions of the fish endocrine system have recently been carefully evaluated in a comprehensive review of the literature prior to 1957. Pituitary, interrenal, thyroid, and perhaps other endocrine tissues are evidently involved in osmoregulation but the mechanisms appear to differ in various species. For example, ***Fundulus heteroclitus*** is unable to survive in fresh water when deprived of an as yet unidentified pituitary hormone, whereas the ability of many other species of euryhaline fish to withstand osmotic stress seems not to be impaired by hypophysectomy. Pickford points out that euryhalinity has evolved in many different groups of fish, and possibly in as many different ways. Present evidence indicates that rather different mechanisms may be expected in the different species.

Hickman (1958) in our laboratory has re-investigated the physiology of the thyroid in a euryhaline fish and, contrary to the findings of several earlier investigators, showed that demands for iodine—and presumably thyroid hormone—are greater in the marine than in the fresh water environment. Since his work also shows that the standard metabolism is greater in the more saline environment, it is suggestive of some calorigenic function for the fish thyroid. This again is contrary to much of the literature.

Hickman's study is based on the starry flounder (*Platichthys stellatus*), a widely distributed euryhaline flatfish which may migrate far into fresh water and live there for considerable periods. Standard metabolism was measured with flounders ranging in size from 4 to 300 g in waters of 0, 25, and 43‰ salinity. Fish were always adapted to the particular salinity prior to the experiments. Oxygen consumption was consistently and significantly greater in the more saline environments, and the effect was proportionately greater in the smaller animals. This same problem has been many times investigated with different species of fish. Hickman's investigation indicates that some of the confusion has been caused by comparisons of fish of different size and fish not fully adapted or incapable of tolerating the salinities under study. Hickman's comparisons are all based on regression lines for the rate of oxygen consumption as a function of body weight. The results leave little doubt that, in the starry flounder, metabolic demands for mineral excretion are greater than those for water filtration.

The thyroid activity of flounders living in waters of different salinity was studied by measuring the rate of uptake of tracer doses of I^{131}. Like many other workers, Hickman found that thyroid iodine uptake was greater in fresh than in sea water. This effect, however, was shown to be due to a goitrogenic effect of fresh water. When flounders were compared in sea water and in fresh water containing the same amount of iodine, thyroid activity was indeed greater in the marine fish. I^{131} was measured in the blood as well as in thyroid tissues of

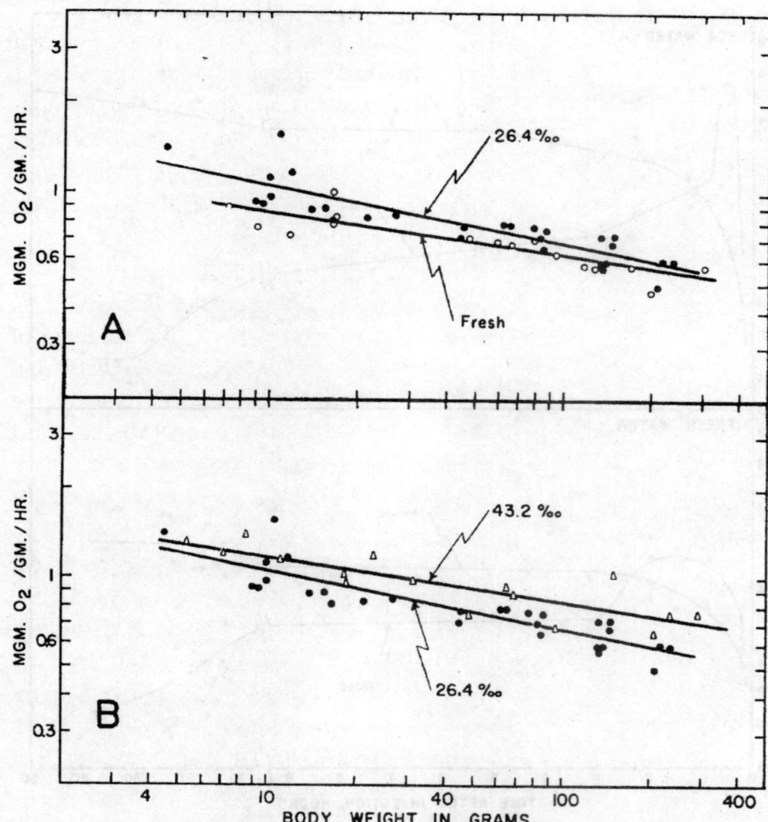

Fig. 11.3. Comparison of the standard metabolic rate of starry flounder in the normal environmental salinity of 25‰ with flounder adapted to fresh water (A) and concentrated sea water of 43‰ (B).

fish maintained in sea water (29‰) and in fresh water containing an amount of iodine equivalent to that of sea water. Iodine was added as a mixture of iodate (80%) and iodide (20%). The difference is apparent both in the absolute amount of radioiodine accumulated and in the rate at which radioiodine is cleared from the blood. It is indicated that osmoregulation in sea water—active excretion of salt—requires more thyroid hormone than osmoregulation in fresh water—primarily a problem of water filtration. The greater oxygen consumption in sea water may or may not be directly related to the thyroid metabolism. These results indicate that thyroid hormone is somewhere involved in the events concerned with general metabolism and/or the regulation of electrolytes.

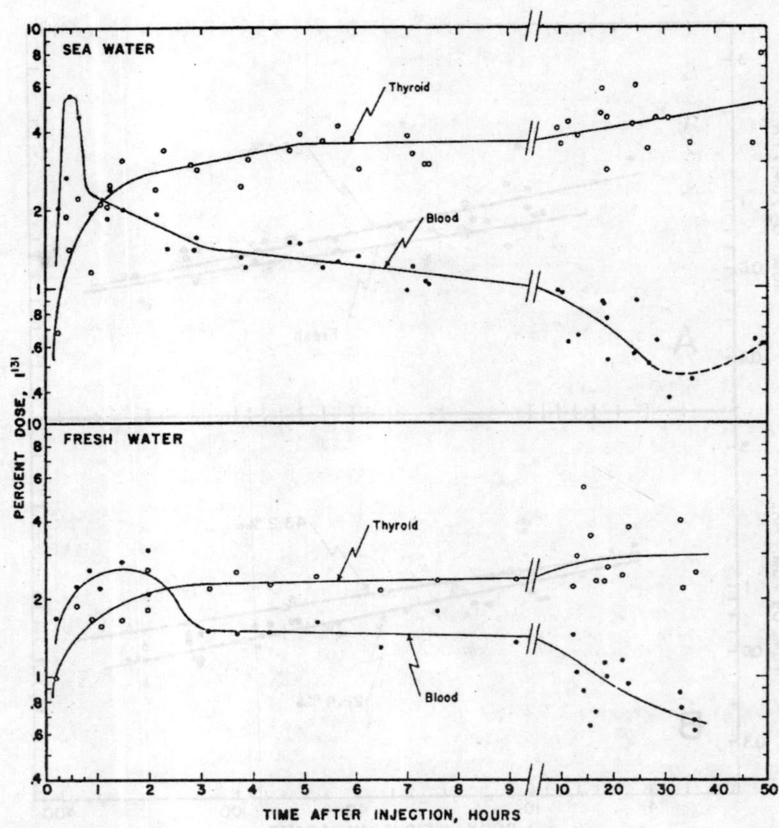

Fig. 11.4. Uptake of I^{131} by the thyroid (optic circles) and the disappearance of I^{131} from the blood (solid circles) of starry flounders adapted to iodine-enriched fresh water and sea water of 29‰.

These studies now permit a more adequate evaluation of the thyroid hyperplasia so frequently described in some anadromous fish and in marine fish retained in waters of reduced salinity. The thyroids in various fish species probably differ greatly in their ability to trap iodine. It might be assumed that strictly marine species and those favouring a marine environment with its high iodine content would be less efficient in trapping iodine from dilute solutions. In waters with little iodine, one trapping mechanism may be adequate, whereas another may fail to accumulate sufficient iodine to produce the hormone required for normal growth and metabolism. Such a difference is probably exemplified in the thyroids of two species of landlocked fish from the Great Lakes. Both were compared with their marine counterparts from the Atlantic coast. The smelt (*Osmerus mordax*) grows to the same size as its marine relative. It has a somewhat more active but by no means hyperplastic thyroid. The alewife (*Pomolobus pseudoharengus*), on the other hand, is only slightly more than half the size of its marine relative and often experiences a spectacular mortality at the time of reproduction. Its thyroid is extremely hyperplastic and may be atrophic. The low iodine content of the Great Lakes waters and their goitrogenicity are well known. Robertson and Chaney (1953) have shown that the assumed increased demands for thyroid hormone at the time of reproduction may induce a marked thyroid hyperplasia in rainbow trout of the Great Lakes region, and that this can be prevented by iodine. Sea-run trout in coastal regions show no such hyperplasia. Limitations in the iodine trapping mechanism may restrict fish in their distribution and success.

The thyroidal hyperplasia described in several species of anadromous fish in fresh water may be goitrogenic (environmental) rather than physiological. Thyroids of pink (*Oncorhynchus gorbuscha*) and chum (*O. keta*) salmon fry retained in fresh water beyond the time of migration were found to be hyperplastic. This was interpreted as a demand for thyroid hormone in connection with the elimination of large amounts of water by salmon physiologically adapted for life in the sea. Later examination of pink and churn salmon thyroids from underyearlings retained for an equally long time in fresh water, but from another region, and on a diet of marine fish, rather than beef liver, failed to show this hyperplasia. Thus, the marked hyperplasia described earlier in thyroids of these species may now be considered partially or wholly due to lack of iodine. This again raises the question as to the significance of the thyroid hypertrophy described in salmon at the time of the smolt transformation. This topic will be considered in the section on migration.

Gaseous Environment

Fish are found in waters which differ greatly both in dissolved oxygen and carbon dioxide, limited oxygen not only restricts the distribution of fish but, at certain levels, can itself control their rate of metabolism. Moreover, carbon dioxide restricts the utilization of oxygen to different degrees in the various species. The endocrine system probably plays a part in the adaptations which fish make to these limiting factors but the nature of its participation is not yet understood. Several lines of study suggest that interesting relationships may be expected. Seasonal changes in oxygen consumption and carbon dioxide sensitivity have been demonstrated in different fish. Seasonal variations, independent of temperature, have been described both in the total metabolism of fish and in some of their isolated tissues. Ekberg (1958) records a consistently lower oxygen consumption by gills of goldfish during the winter months. Gills were studied from goldfish adapted to 10°C and to 30°C. Other groups of poikilotherms likewise show temperature-independent seasonal variations in metabolism. Such findings suggest a photoperiodically controlled mechanism such as that discussed above for temperature resistance.

In this section also, it may be noted that some fish temporarily change their mode of respiration from an aquatic to a terrestrial one, and that characteristic changes in associated endocrinology have been described. Harms' (1929) work on the Indonesian mudskippers (*Periophthalmus*) has often been quoted as evidence for a thyroid regulation of the respiratory physiology. Thyroid treatment induced a premature semi-terrestrial habit in *Periophthalmus* as well as in certain blennies (*Blennius pavo*) which do not normally leave the water. *Periophthalmus* evidently has a particularly active thyroid for a fish and this may be related to its semi-terrestrial habits. It is of interest that the lungfish (*Protopterus annectens*)—also a partial air breather—shows very active synthesis and secretion of thyroid hormone. This activity is considerably reduced during aestivation. Several investigators have been interested in the relationship of thyroid hormone to air-breathing by fishes and a review of their findings is available. The results, however, are by no means conclusive and the detailed endocrinology of these transformations remains to be described.

Light

Fish may live in the bright sunlight near the water surface, or they may live in almost total darkness in the depths of the ocean or in dark caves. Several interesting studies of the endocrinology of cave dwelling species are available. The data suggest that the evolution of fish capable of living continuously in the dark has required profound adaptive changes in the endocrine system. In *Astyanax mexicanus* maintained for long periods in darkness there is a failure in growth and reproduction, correlated with skeletal abnormalities and changes in the endocrine system. The closely related characin, *Anoptichthys jordani*, is blind and lives successfully in caves. Rasquin and Rosenbloom (1954) conclude that the evolution of the blind characin has been dependent on basic changes in endocrine glands as well as other organs such as the photoreceptors.

Physiological colour responses and pigmentary changes associated with the endocrines might also be considered among the adaptations which fish show to different habitats. The physiology and endocrinology of the chromatophores of fish have, however, been discussed in several recent reviews and will not be considered here.

Habitat Selection

Many fish characteristically live within a rather narrow geographical range. Within this restricted habitat they are organized to tolerate normal environmental variations such as temperature and salinity. In addition, they are able to resist to different degrees the unusual extremes in natural conditions. In previous sections, an attempt has been made to show the part which hormones play in some of these physiological adaptations of different species of fish. Other species of fish just as characteristically avoid extremes by swimming for considerable distances from one type of habitat to another. Their geographical range is much greater and, in some cases, extends over thousands of miles. Yet their journeys are restricted to well defined routes and, in a sense, their habitat may be as circumscribed as that of the fish which lives out its life within a few cubic meters of water.

All gradations between these extreme conditions are found. The endocrine system may or may not be involved in a specific way. Trout, for example, often migrate for considerable distances during the summer from warm lake waters into cool streams. Again, some species of juvenile Pacific salmon (genus *Oncorhynchus*) are, as very small fish, characteristically active at night. They swim vigorously at low light intensity and in the fast-flowing Pacific coast streams movement to the estuary or lake is inevitable. It is doubtful whether endocrines play any specific part in these movements. On the other hand, there are other species of Pacific salmon which undergo a marked transformation (smolt transformation) after about one year in fresh water and, in association with this transformation, migrate long distances to salt water. Well-marked endocrine changes are associated with this transformation and it is speculated that the hormones are integrally involved.

Such examples could be multiplied. Perch may show a diurnal migration, feeding in deep lake waters during the day and resting in shallow inshore waters at night. Herring show an annual migration from deeper ocean waters to the shallow inshore spawning areas at the time of reproduction. Both migrations are well-organized mass movements but, in the first case, there is no obvious endocrine association, whereas in the second case the movement is associated with pituitary, thyroid, and gonadal changes. The important point for the present discussion is that the endocrine system is primarily involved in those mass cyclical migrations associated with reproduction or movement from a nursery to a feeding area. The general problem of migration is a behavioural one, which in some cases will be understood only against the background of endocrinology.

This discussion will be confined to two major endocrinological problems associated with these cyclical migrations. The majority, but not all, of these migratory fish move vigorously against water currents in the pre-spawning stages. This is perhaps the most basic general problem in fish migration. Eggs are deposited upstream and the eggs, larvae, or young fish either swim or are carried with the current back to the feeding grounds. This vigorous pre-spawning migration against the current is associated with pituitary, gonadal and thyroid activity. The function of the hormones in this behaviour may be fundamental.

The second of these problems concerns migrations which involve a change in the salinity of the environment. It has been argued that physiological changes, induced by the neuroendocrine system, force the animal to change the tonicity of its environment in order to survive. This theory should be first examined in its more extreme form and an attempt made to evaluate the "compulsion" of habitat change produced by endocrine activity. Further, even though fish may not be compelled to change their habitat in order to survive, they may experience physiological changes which would induce a marked "preference" for a different type of habitat. If the animal were in a gradient of a particular environmental factor (for example, a salinity gradient) it might show a preference in one direction or the other and thus orient its migratory activity.

Upstream Migration

It is difficult to test experimentally the idea that hormones promote upstream swimming. Many devices have been used to measure reactions of fish to current, and in some of these the animals will move either with the current or against it in accordance with their physiological state or the environmental conditions. However, most of the long, and more spectacular, migrations of fish occur in deep open waters where there seems to be a lack of visual or rheotactic stimulation of the type experienced in small experimental streams of water. The mechanism of orientation in deep water has not been explained, and, consequently, an adequate test of the effect of hormones on upstream migration has not been devised.

Thyroxine and gonadal steroids have been tested only on small fish in areas with numerous reference points of both a visual and a rheotactic nature. Young salmon swim more vigorously against water currents after treatment with synthetic thyroxine, testosterone, or stilboestrol. The general unstimulated locomotor activity is also greater. Young salmon (genus *Oncorhynchus*) or goldfish (*Carassius auratus*) swim more in quiet water also when treated 5 to 15 days with these compounds. Furthermore, the intensity of electrical stimulus required to produce a reflex response is lowered. These experiments with fish agree with those on other groups of vertebrates in which thyroxine and the gonadal steroids increase locomotor activity. In our experiments no specific behaviour patterns were induced, and this was perhaps not to be expected. It has been suggested that these hormones are involved in the appetitive behaviour associated with migration, and that the increased locomotor activity associated with them brings the fish into situations where appropriate environmental situations release the genetically specific behaviour of migration. As far as migration is concerned these hormones could potentiate the characteristic activities without being responsible for any specific orientation.

Diadromous Migrations

It is well-known that some anadromous and catadromous fish must move from fresh to salt water or vice versa in order to survive and reproduce. Chum and pink salmon (*Oncorhynchus keta* and *O. gorbuscha*), for example, do not survive longer than a few months as juveniles in fresh water. Coho (*O. kisutch*), on the other hand, may survive to maturity but do not reproduce successfully, whereas sockeye (*O. nerka*) produce successful fresh-water or sea-going populations. There may be a certain "compulsion" to migration in some species. It has not, however, been proven in the examples just cited that the associated endocrinological changes are responsible for this compulsion.

A marked decline in the tissue chlorides of the eel is associated with its transformation prior to seaward migration. On the basis of this demonstration, Fontaine and his colleagues first suggested that such a demineralization, possibly under endocrine influence, would impose physiological restrictions or limitations which could only be counteracted by movement into saline waters. This theory was elaborated for several species of fish by Fontaine and Koch (1950). Its most recent statement is given by Fage and Fontaine (1958). In addition to the eel (*Anguilla anguilla*), a loss in the inorganic constituents also precedes the seaward migration of *Salmo salar*, *Oncorhynchus masou*, and *Salmo gairdneri*. It may be a characteristic feature of all seaward migrations, but this remains to be shown. There is certainly no strict "compulsion" imposed by such a demineralization, since the last three species mentioned may complete their life cycle in fresh water.

There is one good example of an endocrine-induced physiological change which seems to require a movement to waters of different salinity. Stickleback (*Gasterosteus aculeatus*) of the subspecies *gymnurus* can only reproduce successfully in fresh water. As young animals they are remarkably resistant to salinity change but at sexual maturity show a markedly decreased euryhalinity. Thyroid activity is associated with this decreased euryhalinity and thyroid treatment of immature sticklebacks likewise decreases their ability to tolerate sea water. In this case action of thyroid hormone imposes the requirement of a change in habitat. Increased appetitive behaviour associated with thyroid (and gonad) stimulation would probably be inadequate to orient these fish through estuaries and other areas where water currents and other orienting factors are not obvious. Active changes in salinity preference associated with alterations in endocrine activity have been described and are discussed in the following paper.

Finally, it should again be emphasized that fish migration is based on genetically fixed patterns of behaviour which differ in a characteristic way in the different species. In some species changes in hormone production are essential to the change of habitat. This may be particularly true for diadromous fish but is not even characteristic of all diadromous migrations. For many strictly fresh-water or marine species which travel long distances during specific periods of their life history changes in endocrine secretion seem to play an important part in the generalized appetitive behaviour associated with migration, but may not be involved in any more specific manner.

12

HORMONES IN MIGRATION

This paper deals with the role of external factors and changes in pituitary, gonad, and thyroid activity in the timing of migration in the three-spined stickleback (*Gasterosteus aculeatus* L.), with some preliminary information on results of a similar study in juvenile Pacific salmon (*Oncorhynchus*).

MATERIAL AND METHOD

The three-spined stickleback undertakes a prespawning migration from the sea to fresh water during the spring and a postspawning migration from fresh to salt water during the autumn. In Holland the prespawning migration takes place between January and May with the maximum activity in February and March. The breeding season lasts from mid-April until the end of June or the beginning of July, while the postspawning migration commences after the breeding season and continues until November. The offspring also proceed to sea at this time and mature at the end of the first year.

The concept underlying the method used in this study is that migration is preceded by an internal change which induces migration-motivation or disposition (caused by external and/or internal factors). Once the fish are in a migration-disposition state, appropriate external factors are able to activate the migration-drive which induces migratory behaviour. A change in salinity preference of the stickleback has been used as the criterion for the induction of migration-disposition. When animals show a change in preference from salt to fresh water it is assumed that they have come into a physiological condition (disposition) which, under the influence of appropriate external conditions, would induce actual migration from the sea to fresh water. When animals show a change in preference from fresh to salt water it is supposed that migration-disposition would, under the influence of appropriate external conditions, activate migration from fresh water to the sea. It must be emphasized that the change in salinity preference is not supposed to be the primary factor causing the onset of migration. It is assumed to be an accompanying phenomenon, indicating that migration-disposition has been induced.

Salinity tests were carried out in wooden troughs 100 cm long, 15 cm wide, and 13 cm deep. The troughs were divided by glass plates into four watertight compartments of equal size. Two compartments at one end were filled with fresh water, and the two others with sea water. To connect the four compartments a thin layer of fresh water was allowed to flow over the top of the glass plates, enabling the fish to move freely throughout the trough. The salinities used were about twelve parts per thousand in the outermost and about eight parts per thousand in the innermost of the two salt water compartments. In each experiment the fish were tested three times; to avoid interference of position preference with salinity preference the location of the salt and fresh water compartments in the trough was changed for each test. The distribution of the fish in the four compartments was noted each 15 sec during a period of 15 min, or longer. The total number of fish that had been present in the salt water or in the fresh-water halves of the trough were added and the chi-square test used to measure the departure from uniform distribution. In control tests uniform distribution was achieved when the fish were tested in a trough containing only fresh water.

Annual Cycle of Salinity Preference of Sticklebacks Under Natural Conditions

To test the value of the salinity preference technique, sticklebacks from nature, or held in the laboratory under nearly natural conditions, were tested periodically for one year. One series of experiments was made with adult fish and one with under-yearling animals.

The results show that prior to the breeding season maturing fish prefer fresh water. This coincides with the time of spring migration to fresh water. At the end of the breeding season, when they migrate seaward, the sticklebacks prefer salt water. Under-yearling fish (two or more months of age) showed a preference for salt water; this preference corresponds to the time of the seaward migration.

These experiments justify the use of changes in salinity preference as a criterion for the induction of migration-disposition. They also

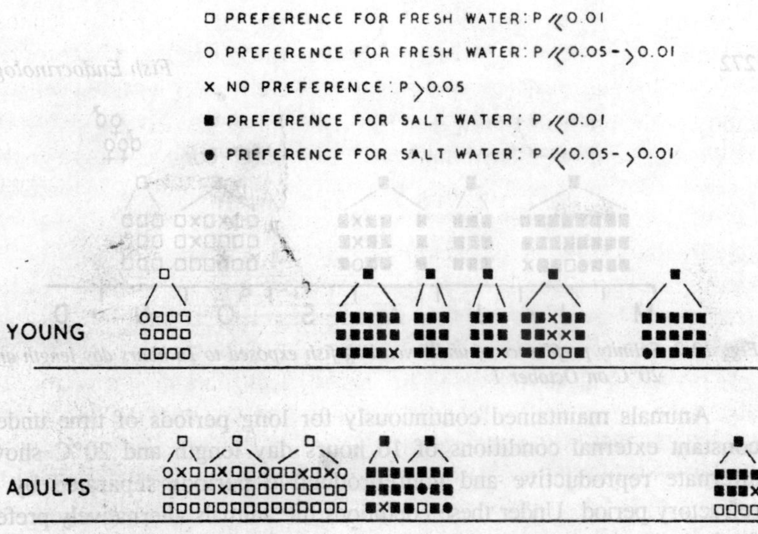

Fig. 12.1. Annual cycle of salinity preference of adult and under-yearling fish living under natural conditions.

show the close relationship between gonadal cycle and cycle in salinity preference under natural conditions.

Gonadal Cycle and Salinity Preference

A series of experiments with controlled photoperiods showed that sexual maturity in sticklebacks could be induced at any time of the year by proper manipulation of the daily photoperiod and temperature. In all cases it was found that full maturation of the gonads was accompanied by a change in preference from salt to fresh water.

The animals used in this experiment were under-yearling fish hatched in the laboratory in April and May and subsequently reared under almost natural conditions. In June, July, and August these fish preferred salt water. On October 1 they were subjected to a day length of 16 hours and a temperature of 20°C. When tested in the second half of October and the first half of November their preference had changed and they preferred fresh water. On the average the five animals matured 49 days after the transfer to long day length and high temperature. Thus, a fresh-water preference can be induced at a season when the animals normally prefer salt water; the external stimulus which elicits this fresh-water preference also causes gonadal maturation.

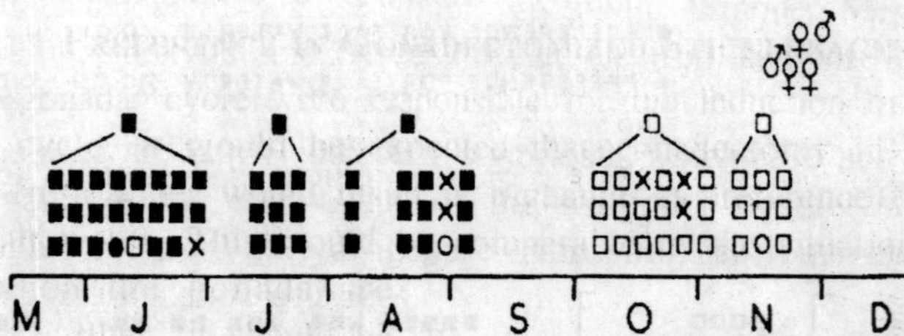

Fig. 12.2. Salinity preference of under-yearling fish exposed to 16 hours day length and 20°C on October 1.

Animals maintained continuously for long periods of time under constant external conditions of 16 hours day length and 20°C show alternate reproductive and nonreproductive periods separated by a refractory period. Under these conditions the animals alternatively prefer fresh and salt water, their particular preference corresponding with maturation and regression of the gonads. It is evident from this figure that during the first two months after the end of the breeding period the fish showed a prevailing salt-water preference. After about three months the salt-water preference became less pronounced and after four months there was a consistent fresh-water preference. On the average the animals matured again five months after the end of the first breeding period. A similar experiment was carried out with female sticklebacks and the result was the same. The experiment stresses the close relationship between gonadal maturation or regression and changes in salinity preference.

Animals maintained continuously under constant external conditions of 8 hours day length and 20°C never attain maturity if such treatment is begun immediately after breeding. Figure 12.3B shows that such animals never change their initial salt-water preference into one for fresh water, even when kept under this condition for over 8 months. Similarly, when young fish 3-4 months old were submitted to 8 hours day length and 20°C they did not attain maturity even after one year, nor did their salt-water preference change. However, when they were transferred alter 14 months to a long day length (16 hours), they matured on the average 26 days after the transfer. The fish also showed a change in preference from salt to fresh water in this period.

It can be concluded from the above experiments that attainment of gonadal maturity coincides with a change in preference from salt to fresh water, whereas gonad regression is correlated with a change in preference from fresh to salt water. The same external factors causing gonad development also induce a change in preference from salt to fresh water; in both cases the external factors probably produce their effect by way of the pituitary. As no maturation or change in preference from salt to fresh water occurs under certain constant external conditions (short day length and high temperature), it is evident that there is no intrinsic rhythm inducing these changes; they are dependent on the presence of certain external factors such as a long day length. On the other hand, gonad regression and a change in preference from fresh to salt are caused by intrinsic mechanisms as these changes occur under constant external conditions. Although the pituitary is probably involved in both intrinsic systems, the experiments do not show that these mechanisms are the same.

It must be pointed out that this kind of experiment is not suited to prove a possible cause-and-effect relationship between gonadal cycle and migration. It has been shown that there is a close relationship between the two phenomena, but it has not been shown that migration is caused by gonad changes. One of the ways to investigate the nature of the relationship is to gonadectomize animals and study its effect on migration. Experiments of this kind will be described in the next section.

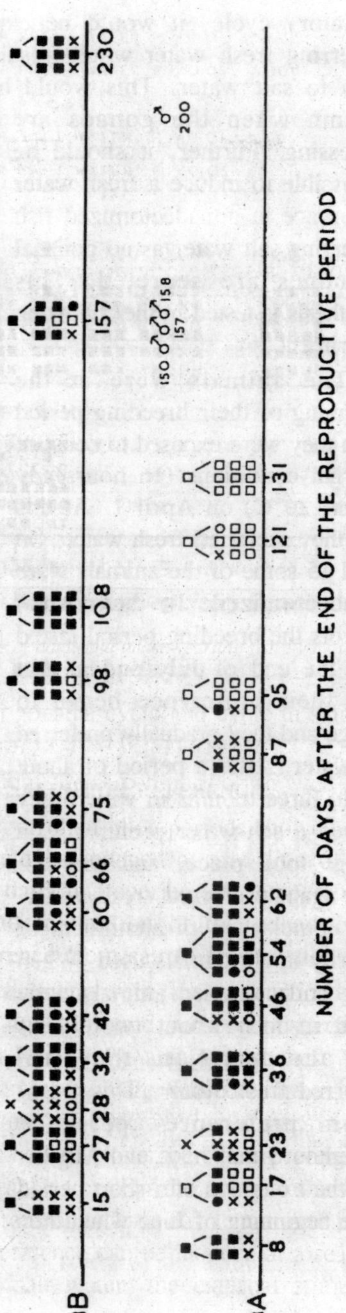

Fig. 12.3. Salinity preference of fish tested at various times after the end of the reproductive period. A, exposed to a constant day length of 16 hours and 20°C; B, subjected to a constant day length of 8 hours and 20°C.

Salinity Preference in Gonadectomized Sticklebacks

If the gonadal cycle were responsible for the induction of the migratory cycle, it would be expected that gonadectomy in fish preferring fresh water would result in a change in preference from fresh to salt water. This would be comparable to the situation in autumn when the gonads are regressing. Further, it should be impossible to induce a fresh-water preference in gonadectomized fish preferring salt water as no gonadal hormones are secreted. This hypothesis is tested in the following experiment.

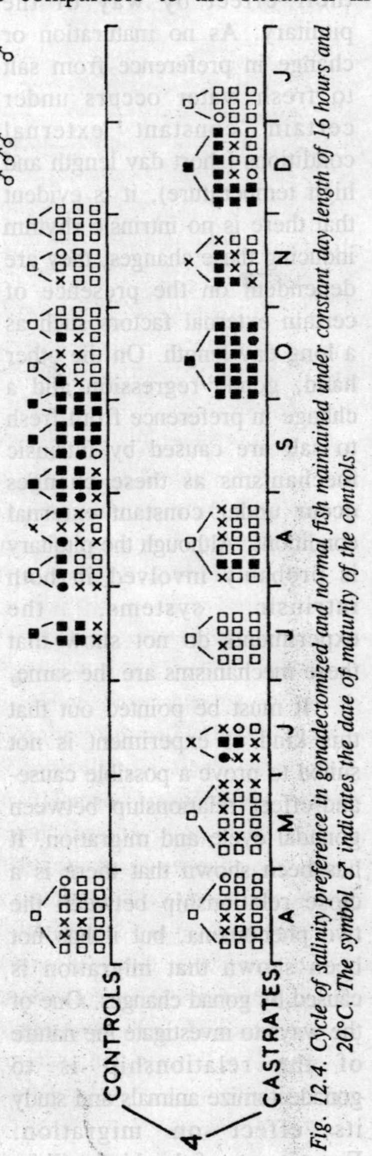

Fig. 12.4. Cycle of salinity preference in gonadectomized and normal fish maintained under constant day length of 16 hours and 20°C. The symbol ♂ indicates the date of maturity of the controls.

The animals were at the beginning of their breeding period when they were exposed to constant external conditions (16 hours day length; 20°C) on April 1. At this time they preferred fresh water. On April 15 some of the animals were gonadectomized. In the normal controls the breeding period lasted until the end of July and at that time their preference began to change and they gradually preferred salt water. After a period of about two to three months in which they showed a salt-water preference, a change took place, and the fish once again preferred fresh water. This coincided with attainment of maturity in all animals used. When the gonadectomized animals were tested in May about two weeks after the operation, they still preferred fresh water. This fresh-water preference persisted throughout June, July, and August, with the exception of a short period at the beginning of June when they had no preference. However, at the beginning of October they changed their preference and preferred salt water. This is considerably later than the corresponding change in preference in the controls. The salt-water preference lasted until December when it gradually began to change. At the end of this month and the beginning of January the fish once again showed a consistent fresh-water preference. This change also took place later than in the controls.

Two main conclusions can be drawn. In the first place, fish initially preferring fresh water showed no immediate change in preference after gonadectomy. The change in preference from fresh to salt did eventually take place, but as the change occurred later in the operated than in the normal fish, it cannot be related to gonadectomy. This result is not in agreement with the hypothesis that gonad regression, and thus a low level of gonadal hormones, would be responsible for the change in preference from fresh to salt water, i.e., for seaward migration. In the second place, a change in preference from salt to fresh water—hence the induction of spring migration—is not primarily due to an increase in the level of gonadal hormones, as a similar change can be induced in gonadectomized animals by proper manipulation of the daily photoperiod and temperature.

Although the experiments give conclusive evidence that gonadal development or regression are not the primary cause of stickleback migration, there is some evidence that they may modify the time at which this condition is induced. This can be derived from the facts that (a) the change in preference from fresh to salt water, and salt to fresh water, takes place later in the gonadectomized than in the normal animals and (b) that in some cases there is a slight tendency for the fish to change their preference for a short period after gonadectomy.

In birds, which are the only other migratory animals in which the effect of gonadectomy has been studied, it was also found that the gonads do not seem to play an essential role in the causation of migration although more data are needed before a definite conclusion can be drawn.

Influence of Thyroxine and Thiourea on Salinity Preference

Many investigators of bird and fish migration are of the opinion that changes in thyroid activity may initiate migratory behaviour because the thyroid gland shows increased activity at the time of migration. In the stickleback, Koch and Heuts (1942) were able to show that feeding of thyroid substance resulted in a decreased ability of osmoregulation of fish living in sea water. Moreover, in species of fish migrating between the sea and fresh water, migration is often correlated with changes in osmoregulation. Therefore, it was decided to study the effect of thyroxine and thiourea (thyroid inhibitor) on salinity preference in the stickleback.

In an experiment, started on October 17, five animals were placed in a solution (which was renewed each day) of thyroxine 1 : 1,000,000. On October 21 this concentration was doubled. The fish showed an initial preference for salt water, but on the fifth day of the treatment they had completely changed their preference, this new preference continuing for three days. The controls, on the contrary, maintained their initial salt-water preference during this period. This experiment shows clearly that thyroxine affects salinity preference and induces a fresh-water preference in fish initially preferring salt water.

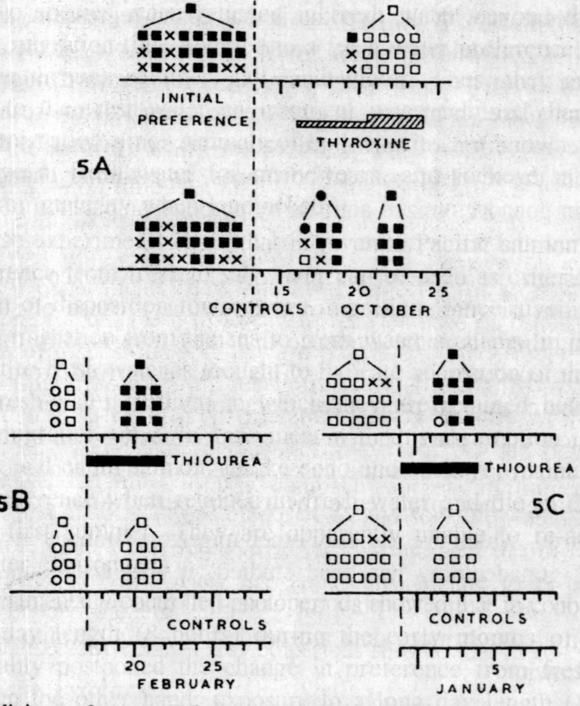

Fig. 12.5. Salinity preference of fish treated with thyroxine or thiourea. A, salinity preference of fish initially preferring salt water and treated with thyroxine; B, salinity preference of fish initially preferring fresh water and treated with thiourea; C, salinity preference of gonadectomized fish initially preferring fresh water and treated with thiourea.

Adult fish preferring fresh water were used to test the effect of thiourea on salinity preference. On February 17 they were placed in a solution containing 0.072 per cent thiourea. During the experiment the controls maintained their preference for fresh water. However, the thiourea group, when tested for the first time on the third day, had already changed their preference and now preferred salt water. This preference was maintained during the next two days. Thus, it is evident that thiourea is able to induce a salt-water preference in animals initially preferring fresh water. The average duration of the latency periods of thyroxine was 3-5 days and of thiourea 3-4 days.

In these cases also, the effect is not brought about by intermediation of the gonads. As in normal animals, thiourea is able to induce a salt-water preference in gonadectomized fish initially preferring fresh water. It is possible, as suggested by the work of Koch and Heuts (1942), that thyroxine affects salinity preference in sticklebacks by disturbing the osmotic system. Effects of thiourea on mineral metabolism have not yet been studied. Possibly, a low level of thyroid hormone in the blood produced by this anti-thyroid material disturbs the osmotic system of animals adapted to life in fresh water, and thus causes them to prefer salt water.

Another possibility is that the influence of thyroid hormone in high or low concentration on osmoregulation is not direct but stimulates or inhibits the secretion of other hormones (e.g., interrenal hormones, growth hormone), which in their turn affect osmoregulation.

Further, thyroid hormone in the blood may exert a direct influence on the central nervous system, causing a change in the behaviour pattern responsible for salinity preference or even in migratory behaviour itself. In this case a disturbance in the osmotic equilibrium would not be the primary cause of the change in behaviour, but a secondary effect. As a secondary effect it may influence migration in that it causes stress as the animals are no longer adapted to their environment.

Salinity Preference, Thyroid Activity and Migration in Juvenile Pacific Salmon (*Oncorhynchus*)

Salinity preference in four species of juvenile salmon, namely chum (*O. keta*), pink (*O. gorbuscha*), coho (*O. kisutch*), and sockeye (*O. nerka*) has been studied. The results of an initial study by Houston (1957) showed that juveniles of several species of *Oncorhynchus* responded positively to sea water at the time of seaward migration. These findings were confirmed and worked, out in more detail by the present author.

Chum and pink salmon migrate to the ocean as fry after hatching in the early months of the year. After a brief period of fresh-water preference following hatching, they preferred salt water and maintained this preference throughout the summer. They die during the summer if retained in fresh water. Coho and sockeye salmon, on the other hand, usually migrate seaward as smolts when they are one year of age. Occasionally sockeye, as very young fry, and coho, while still under-yearling, may migrate to the ocean, or they may remain in fresh water for two years before migrating. These species as very young fry showed a short period of salt-water preference, but during summer this preference changed and they preferred fresh water. This preference persisted through fall and winter until spring of the next year when they underwent the parr-smolt transformation. At this time their preference changed and they preferred salt water. This change corresponds to the time of the onset of seaward migration. Smolts of these two species when retained in fresh water beyond the normal seaward migration time reverted to a fresh-water preference, this state persisting through summer, fall, and winter. Thus, they are able to readapt to the fresh-water environment when retained there beyond normal migration time. Animals of both species are known to be able to survive in fresh water for many years, and sockeye salmon may even attain maturity when landlocked.

These experiments show that in juvenile Pacific salmon a change in preference from fresh to salt water can be used as criterion for the induction of disposition for seaward migration. Since juvenile salmon show no migration from the sea to fresh water a change in preference from salt to fresh water is thought to indicate adaptation of the animals to the fresh-water medium in which they are retained beyond their normal migration time. In agreement with this assumption is the fact that pink and chum salmon, unlike coho and sockeye, maintain a salt-water preference when retained in fresh water and die in the course of their first summer. They are apparently unable to re-adapt to a fresh-water environment.

Experimentally controlled photoperiods showed that in coho yearlings a short day length (8 hours) during the early months of the year considerably postponed the change in preference from fresh to salt water. On the other hand, exposure to a long day length (16 hours) induced an earlier change in preference than in fish kept under normal day length conditions. The temperature in all experiments was the same (between 5° and 8°C). This indicates that day length influences the time of induction of migration-disposition and thus of seaward migration—probably through affecting pituitary activity.

In juvenile salmon, thyroxine and thiourea influence salinity preference. In general however, the results have been inconsistent and more experiments are required before a more definite statement can be given on the role of thyroid hormone in the induction of migration-disposition in juvenile salmon.

Thyroid activity at the time of seaward migration was determined by means of uptake of radioiodine. The results so far seem to indicate that prior to the migratory period, when the animals still prefer fresh water, thyroid activity is lower than at the peak of migration when they prefer salt water. This suggests that thyroid activity may be involved in the onset of seaward migration, although it should be kept in mind that the increase in thyroid activity may be the consequence of the fact that the animals are in an environment to which they are no longer adapted.

Conclusion and Discussion

The following theory is proposed for the causation of migration in the three-spined stickleback. In fall and winter, as a consequence of the short day length and low temperature, the pituitary-thyroid system is relatively inactive, and as a consequence the animals prefer salt water. In nature they winter in the sea. In the early months of the year the increase in day length and temperature increases the activity of the pituitary-thyroid system and brings the fish into migration-disposition—indicated by a change in preference from salt to fresh water. When in this condition external factors are thought to induce actual migration to fresh water. At the same time the pituitary-gonad system is activated and the gonads start to mature. When the fish reach fresh water they are ready to breed. The breeding season is terminated by an intrinsic mechanism causing a decrease in activity of the pituitary-gonad system. At the same time an intrinsic mechanism also causes the pituitary-thyroid system to become less active with the result that the animals come into migration-disposition, indicated by the change in preference from fresh to salt water. Again external factors are thought to induce actual migratory behaviour.

Data obtained so far in the study on migration in juvenile Pacific salmon have led to the following tentative hypothesis. Like in the sticklebacks, external conditions affect the time at which a change in preference from fresh to salt water takes place and thus also influence the induction of migration-disposition. The external factors most probably act by influencing pituitary activity, which in turn affects the activity of the thyroid gland. It is thought that in juvenile salmon, as in the stickleback, the activity of the thyroid gland may be a primary cause of migration.

Many species of fish at the time of migration show changes in activity of pituitary, thyroid, and interrenal glands, and changes in metabolism and general activity. Changes in pituitary and thyroid activity, and in metabolism and general activity have also been observed in birds at the time of migration. These findings led to the general hypothesis that migration is caused by changes in the neuroendocrine system, induced by external and intrinsic factors. However, it is as yet impossible to give a more detailed picture of the way in which migration is caused in the various species of fish and birds. It must be kept in mind that a complicated behaviour pattern such as migration most probably is not caused by one factor but is induced by intricate interaction between a multitude of extrinsic and intrinsic factors, which may differ in the various species of migratory animals.

INDEX